IISS

STRATEGIC SURVEY 1996/97

Published by  for

The International Institute for Strategic Studies
23 Tavistock Street
London WC2E 7NQ

Strategic Survey 1996/97

Published by Oxford University Press for
The International Institute for Strategic Studies
23 Tavistock Street, London WC2E 7NQ

Director Dr John Chipman
Editor Sidney Bearman

This publication has been prepared by the Director of the Institute and his Staff, who accept full responsibility for its contents, which describe and analyse events up to 26 March 1997. These do not, and indeed cannot, represent a consensus of views among the world-wide membership of the Institute as a whole.

Managing Editor Rachel Neaman
Production Supervisor Mark Taylor

First published April 1997

ISBN 0 19 829296 1
ISSN 0459 7230

Strategic Survey (ISSN 0459 7230) is published annually by Oxford University Press.

The 1997 annual subscription rate is: UK £25.00; overseas $US39.00.

Payment is required with all orders and subscriptions. Prices include air-speeded delivery to Australia, Canada, India, Japan, New Zealand and the USA. Delivery elsewhere is by surface mail. Air-mail rates are available on request. Payment may be made by cheque or Eurocheque (payable to Oxford University Press), National Girobank (account 500 1056), credit card (Access, Mastercard, Visa, American Express, Diner's Club), direct debit (please send for details) or UNESCO coupons. Bankers: Barclays Bank plc. PO Box 333, Oxford, UK, code 20-65-18, account 00715654.

Claims for non-receipt must be made within four months of dispatch/order (whichever is later).

Please send subscription orders to the Journals Subscription Department, Oxford University Press, Great Clarendon Street, Oxford, OX2 6DP, UK. Tel: +44 (0) 1865 267907. Fax: +44 (0) 1865 267485. e-mail: jnlorders@oup.co.uk

Strategic Survey is distributed by M.A.I.L. America, 2323 Randolph Avenue, Avenel, New Jersey, NJ 07001, USA. Periodical postage paid at Newark, New Jersey, USA and additional entry points.

US POSTMASTER: Send address corrections to *Strategic Survey*, c/o M.A.I.L. America, 2323 Randolph Avenue, Avenel, New Jersey, NJ 07001, USA.

PRINTED IN THE UK by Bell & Bain Ltd, Glasgow.

Contents

List of tables and figures

Perspectives

Throughout the world in 1996, foreign and security affairs were driven by a pervasive and persistent parochialism. Governments found themselves preoccupied with domestic considerations. Not only was their attention firmly fixed inwards, but for the most part internal stresses and tensions prevented them from taking actions that they almost certainly would have taken had their hands not been tied. This tendency to look inwards restricted the potential for multilateralism.

Domestic concerns have always influenced foreign policy. What characterised 1996, however, was the ubiquity of the phenomenon. National elections in such significant countries as France, Israel, Japan, Russia, Taiwan and the United States impinged on foreign-policy choices, while in China, the imminent death of the 92-year-old patriarch Deng Xiaoping focused the leadership's attention on the struggle for power. In their search for votes, or for support, leaders vying for the top position adjusted their policies to reflect what they thought their constituents wanted. Almost inevitably, this did not include foreign issues.

In the past, a similar spate of elections has not always produced such domestically driven policies. For Western countries, the present absence of a formal external threat reinforces their introverted tendencies. Notwithstanding the unsettled nature of the contemporary world, a feeling of security permeates the major industrialised nations. The conflicts that rage in many parts of the developing world are localised, internal conflagrations – and while a humanitarian impulse can sometimes move states to intervene, this rarely lasts long enough to create support for lengthy foreign entanglements. The Cold War tendency to see core interests indirectly at stake in distant parts of the globe has now totally eroded. Even the brief post-Cold War sense of 'humanitarian obligation' has begun to give way to colder *realpolitik* calculations of what can be done.

Comfortably cocooned in their own sense of security, citizens of democratic countries are in no mood to sacrifice their well-being for supposed international advantage, nor to rally to the service of a purely humanitarian goal. No US President today would dream of calling upon his people, as President John F. Kennedy did in January 1961, to 'pay any price, bear any burden, meet any hardship ... to assure the survival of liberty'. He would be met with scoffs of derision, not nods of assent.

Leaders, therefore, when contemplating actions outside their own countries balance the possible gains against the impact such intervention will have on their own standing at home. Some actions, of course, are neutral in this respect, have little cost, and can be undertaken without trepidation. Others are undertaken for their impact on the domestic affairs of other countries. For, if it is true that domestic politics ruled foreign policy more than usual in 1996, it has also been the case, especially for the US, that such foreign policy as is developed concerns the domestic policies of other states. The policies towards 'rogue states' have in some measure been developed in the hope of influencing their domestic politics. And certainly the policies directed towards Russia and the states of Eastern and Central Europe (in the context of NATO enlargement) have been crafted to produce positive effects on the domestic politics of the target states.

Foreign policy, like domestic policy, now has a twin meaning, and this is perhaps one of the genuinely new characteristics of the post-Cold War world: in some way or other, domestic politics has become the concern of other countries, even if that concern does not always translate into action.

The United States Leads the Way

The beginning of the 1995 Congressional session set the scene for a political stand-off in the United States. Not only had the electorate returned a Republican majority, but that majority had a radical domestic programme that reduced the President's role in domestic legislation to that of an opposition voice armed with a veto. President Clinton turned instead to foreign policy as an arena in which he could make a positive impact. Yet, even in this sphere, he found it necessary to placate both his Republican adversaries in Congress and the conservative-minded US electorate. The struggle for domestic position and electoral favour, particularly in an election year, impinged heavily on the creation and implementation of US foreign policy.

US efforts to punish foreign companies with business dealings with Cuba and Iran or Libya epitomised this trend. The eponymous Helms–Burton Act directed against Cuba, and the similar D'Amato Act directed against Iran and Libya, established criteria for US action which NATO countries and Japan immediately condemned as an unacceptable exercise of 'extra-territorial' legislation. Clinton felt constrained to sign this ideologically driven legislation, particularly the D'Amato Act, because of the proximity of the election, even though he recognised that it would hamper his pursuit of other foreign-policy goals. In an effort to smooth the way forward he has twice, as allowed by law, postponed exercising the more egregious aspects of the Helms–Burton Act.

This US unilateral action underscored just how far the international community has to go before effective multilateralism can be considered anything but a pretence. The offended states claimed in the World Trade Organisation (WTO) that the US action restricted trade. But the Clinton administration, despite the fact that it had lobbied hard for the WTO and had hailed its establishment as one of its proudest foreign-policy achievements in 1995, immediately declared its rogue-state legislation a matter of 'national security' and thus not within the jurisdiction of the WTO. Before the WTO had time to prove its worth, the US provided an escape clause for any other country embarrassed by a similar action in the future, whether the WTO accepts the US position or rules against it, for Washington is then likely simply to disregard the WTO ruling.

The hopes for a world where multilateral concepts have priority over parochial domestic factionalism were dented by the US refusal to grant UN Secretary-General Boutros Boutros-Ghali a second term in office. The lone negative vote that the administration initially cast in the face of an otherwise unanimous agreement for a renewed term was inspired less by a belief that Boutros-Ghali was a poor choice than by a sense that the head of the UN should pay the price for US Congressional disappointment with the organisation's accomplishments in the past few years. The United Nations is not a favourite of many Americans, and to many Republicans in Congress it is a bloated, worthless organisation whose main purpose is to suck money for dubious international adventures out of the US taxpayer.

Senator Jesse Helms of North Carolina, ensconced after the Republican sweep in 1994 as Chairman of the Senate Foreign Relations Committee, has perhaps the least complimentary views of the UN of any in the Senate. Helms particularly held Boutros-Ghali responsible for not carrying out reforms that he had long demanded – demands that would shut down many of the UN's present activities. Boutros-Ghali's attempts at reform, not always whole-hearted, had been shackled by a lack of funds. At the end of 1996, the United States was $1.2 billion in arrears to the UN, and control over payment was not in the hands of the administration, but in Helms' clenched fist. When Kofi Annan, the US-backed winning candidate for Secretary-General, came to Washington to discuss problems with new Secretary of State Madeleine Albright, she asked him to discuss with Senator Helms the reforms he might carry out. The display of where the balance of power lies was unmistakable.

Other US foreign-policy initiatives were not as blatantly dependent on domestic political considerations as these, but they gathered support from parochial concerns. President Clinton's decision to back early NATO enlargement probably owed most to his – and his advisers' – belief in the efficacy of Wilsonian democratic ideals. Ensuring the security of some of the more economically and politically advanced new European nations would do much to help solidify their still-nascent democracies. But that Clinton delivered a speech in Detroit just before the election announcing an accelerated timetable for NATO enlargement was undoubtedly influenced by the existence of 20 million voters of Central European origin located primarily in important electoral states like Ohio, Michigan and Illinois.

Similarly, the US decision to confront China's belligerent actions against Taiwan in the run-up to the Taiwanese elections in March 1996 was undoubtedly taken for sound balance-of-power reasons. Indeed, sending two aircraft carrier battle groups to the area not only made US concerns palpable, but also reassured other friendly nations in the Asia-Pacific which constantly fear that the US might withdraw its support and influence, as well as its troops, in the not-too-distant future. Yet this action also played to the US electorate where Taiwan is now considered a sharp democratic contrast to China. And the administration's stalwart efforts to steady the Middle East peace process when it faltered in 1996 also resonated loudly with the large majority in the United States who feel strongly that this is among the most important areas for US diplomatic engagement.

Even though 1997 is not an election year in the United States, domestic political realities can be expected to suffuse foreign-policy considerations – at the least as a constant drumming in the background, and at the most as the determinant of action. It is a mistake to think that because the sitting President cannot stand for election again he is free of political restraints. Should he take actions against the electorate's wishes, his position would weaken and his influence would ebb. Congress, particularly when in the grip of the opposition as in 1996–98, would rush to fill the developing power vacuum. In addition, while Clinton cannot be the Democratic Party nominee again, actions that the administration takes will deeply affect the viability of Vice-President Al Gore, Clinton's own choice as successor.

Any US President in the last two years of his term, if he cannot run again, is a 'lame duck'. For President Clinton, however, the effects of the fund-raising scandal that dominated the first months of his new term in office – particularly as it involved accepting money from foreign sources – have made his position even more delicate much earlier. Inevitably, his policies regarding China, Taiwan and other Asian nations will be viewed through the prism of these collections. The fact that Clinton cannot run for office again does not free him from the need to pay strict attention to the exigencies of domestic political infighting while forming his foreign policies – in some regards, it may even make him more vulnerable to their dynamic.

Towards a Single European Parochialism

While US provincialism has had a long tradition, European parochialism took on a new mantle in 1996. The key powers in the European Union were not so much driven by national events as by the wider struggle to create a single Europe. While the 'under construction' signs were up, the Europeans barely lifted their gaze above their own ramparts.

To be fair to the EU member-states, their building project was behind schedule and the workers were increasingly insecure about their jobs. Most EU countries (with the striking exception of Ireland, the Netherlands and the UK) were barely out of economic recession and yet were telling their people that they had to suffer further hardship in order to meet the criteria laid down by the 1992 Maastricht Treaty on European Union for a single European currency. In truth, the need to reform the EU states' over-burdened social programmes and liberalise employment systems was so stark that even without the Maastricht excuse, harsh reforms would have been necessary.

Whatever the cause, reform was sufficiently unpalatable for most EU countries to focus on finding domestic compromises to sustain reform and the governments in power. In Germany, the task was especially difficult because the population had grown accustomed to sustained growth without tough choices. The price of reunification and the need to remain competitive in increasingly global industries meant that Germany was set for a prolonged process of reform. France, Italy and various smaller European states faced similar social and economic challenges and all seemed to have their jaws locked into the same rhetoric about the need to meet the Maastricht criteria. The French, in particular, could barely contain their frustration, but the government nevertheless struggled on with reform and European union.

The UK's version of parochialism was the result of a tired government running out of political fuel. The ruling Conservative Party's divisions on Europe were no deeper than they were in the opposition Labour Party, but the Conservatives had lost that ethereal, but no less essential, will to govern. The UK's severe doubts about a federal Europe meant that even if the Labour Party should win the elections on 1 May, its EU partners can only expect a change in tone, not in tune.

As far as German Chancellor Helmut Kohl was concerned, the tendency towards Euro-pessimism had to be resisted because of the pressing need to make progress on European union. This required success both for the single-currency initiative and for NATO expansion. Just as Germany was beginning to feel that it was all right to think in terms of German national interest, it could not quite trust itself to do so without being bound into a wider and deeper Europe.

While the goals of a wider and deeper Europe were admirable, they were being pursued in a perverse fashion. Few would dispute that it made more sense to offer the Central Europeans the early prospect of membership in the EU before NATO. But the cost of EU enlargement (for example, financial aid and extending the Common Agricultural Policy) would fall on West Europeans already struggling with painful economic reform. Thus Euro-parochialism dictated that NATO expansion would be reluctantly supported as a second-best method of integrating the Central Europeans into the European family. The price was prolonged economic pain for Central Europe and an unnecessary political and security debate with Russia about NATO expansion. The challenge of monetary union is such that the trend of parochialism is unlikely to change soon: if Economic and

Monetary Union (EMU) is successful, it will only be because of a sustained concentration on the domestic economy. If it fails, the EU states will search for another form of glue for the 'Europe project'.

Given such EU self-absorption, it was not surprising that a more economically constrained and politically unstable Russia saw the process of NATO expansion through domestic lenses. The June 1996 elections returned President Boris Yeltsin to power, but his serious ill-health ensured that the year was dogged by speculation of leadership change and factional politics. As the reformist clans battled prematurely for the Yeltsin succession, the Communist Party nationalists and heirs fumbled for advantage without much success. Under such circumstances the regions had greater freedom of manoeuvre, the war in Chechnya was lost, and Russian foreign policy was little more than creative obstructionism. Given these domestic uncertainties, and the public mood, it was a near certainty that Russia would oppose NATO expansion until it had no choice but to accept it. Russia spoke of its 'strategic partnership' with China, but in truth there was no strategy and little partnership in the relationship. If there was any Russian consolation in this sorry picture, it was that its overwhelming focus on domestic affairs was in tune with the mood of the EU and the United States.

Somersaults in the Middle East

Binyamin Netanyahu is not a good acrobat. An accomplished acrobat can somersault with grace, twirling in the air but landing in the same place without a wobble. Netanyahu's somersaults have been graceless and wobbly. He was elected on the narrowest of margins by a country that wanted a peace process, but one with greater security. Netanyahu began his term as Israeli Prime Minister by slowing the momentum of the process begun by the former Labour government, but in the end accepted that there was greater support at home and abroad for striking a deal on Hebron with the Palestinian Authority (PA) even if on minimalist terms. But even Netanyahu's apparent inclination to slow the peace process where possible, and accelerate it only for cynical reasons – as in his March 1997 announcement that final-status talks could be completed in six months – inspired near-universal pessimism.

Netanyahu wavered on domestic issues as much as on foreign policy. Most notable was the scandal concerning the abortive appointment of Roni Bar-on as Attorney General in January 1997. As the scandal spread and looked as if it would affect the Prime Minister himself, Netanyahu's authority was undermined. Trying to keep his balance, Netanyahu moved sharply to the right to ensure the continued support of the religious ideologues and their parties. Predictably, these factional manoeuvres undermined the prospects for peace and threw the whole process into greater confusion.

The pattern of factionalism-as-usual in Israel made Palestinian politics seem organised by comparison. Chairman Yasser Arafat's position as President of the Palestinian Authority has long been precariously predicated on his ability to deliver Israeli compliance with the peace process. When Netanyahu looked unlikely to deliver his part of the bargain, Arafat's prestige suffered. The Hebron deal was a major gain for Arafat, but not for long. In March 1997, when the *Likud* cabinet defied Palestinian and foreign pressure by deciding to build a new housing development in southern Jerusalem at Har Homa, Netanyahu's commitment to the peace process, and Arafat's credibility, were called into question. It is hard to see anything but an endless stream of such crises until more of the Israeli electorate accepts that only a final peace with the Palestinians will bring greater stability, and insists on a government that will act accordingly.

Given these fruitless diplomatic acrobatics, it was not surprising that little else moved forward in Middle East diplomacy. Most notably, the Israel–Syria dialogue produced no sound, and virtually nothing was heard of the much-touted prospects for new economic cooperation in the region. Various external actors tried their hand at regional diplomacy, but with little impact. France, and to some extent the wider EU, was welcomed by some Arabs, but because Israel remained sceptical, their efforts had no impact on Israel's disputes with Lebanon, Syria or the PA.

The United States remained the only external power with sufficient influence to help create movement, but, in part because 1996 was a US election year, President Clinton, a Democrat with a large Jewish vote, was unwilling to risk a major row with Israel. More importantly, Clinton recognised that as a fellow democracy, and one with a deeply divided electorate, Israel would not respond well to being forced into a peace deal with Arafat. While the time for more pressure may come at the end of a second Clinton term, in 1996 the time was not yet ripe. For all his graceless wobbling, Netanyahu had not yet fallen over and the peace process, though ailing, was still breathing, if weakly. But it would need constant care before it could get back on its feet and move forward.

A Partial Exception in Asia

Parochialism was also a problem in Asia, but because it was accompanied by a palpable sense that a new balance of power was developing in the region, there was a potentially volatile mixture of parochialism and paranoia. The most important arena and most potent paranoia were found in Asia's largest power – China.

Even the clinically paranoid have real reasons for concern. China's enormous experiment with economic reform has been accompanied by enough social unrest, political uncertainty and economic challenge to justify the leadership's concern that China was unstable and that foreigners might seek to exploit such instability. When Deng Xiaoping, the avatar of reform and paramount leader for almost 20 years, died in February 1997, China lost a unifying force. The calming words of Chinese officials that a succession with Jiang Zemin as the 'core leader' had already happened could not mask clear signs that a leadership struggle was under way.

Even without the unsettling possibility that within the region's giant policy-makers would form their policies with succession considerations in mind, there was no reason for complacency about stability. In reality, the region had far more reason to be concerned about the sustainability of reform and turbo-charged economic growth. Since the Japanese economy stalled in the early 1990s, China had taken over as the most dynamic unit of regional economic growth. But it was a common cause of concern in East Asia that Chinese growth was based on a fragile coalition of reformers who counted on rapid prosperity to buy off disaffected and disadvantaged groups in an increasingly decentralised system. With rampant corruption and mass internal migration, the success of Chinese economic reform sometimes seemed like levitation. Now that Deng has died, Chinese leaders will probably struggle, albeit in slow motion, for new ways to meet the challenge of reform.

China's rapid return to 'great powerdom' raised contradictory problems. If Chinese reforms stalled there would be a risk of a regional economic crisis and maybe even mass migration. But further double-digit growth in recent years made it likely that China would emerge as the overwhelming regional power and the only near-peer competitor of the United States. As neighbours contemplated Chinese domestic success, they noticed a creeping assertiveness in Beijing's use of its new muscle.

Nowhere was the musculature more noticed than in Japan. In 1996, 22% more Japanese now identified China as a threat. Beijing's off-hand dismissal of Japanese complaints about Chinese nuclear tests, its pursuit of irredentism in the South China Sea, its naked threats to Taiwan, and the revitalised dispute over the Senkaku/Diaoyu islands, all suggested that as a rising power China was becoming more of a problem.

A more confident Japan might have been able to shrug off such fears and put its faith in a booming economy. But, despite the restoration of a Liberal Democratic Party (LDP) government in 1996, Japan could not return to its previous bravado about 'managing China'. Japan's economy had stagnated, it was in the midst of a difficult campaign to de-regulate its economy and society, and its political system was moving hesitantly towards more far-reaching reform.

Rather than growing more independent, as it desired, Japan in the late 1990s felt the need for even closer relations with the United States, and perhaps with Russia. Because it lived in a precarious neighbourhood with a rapidly growing strongman, Japan no longer had the luxury of introspection. The failure of Japanese Prime Minister Ryutaro Hashimoto to interest Association of South-East Asian Nations (ASEAN) states in closer security relations suggested that Japan would have to rely on an increasingly closer (albeit modernised) alliance with the United States. Other East Asians, it seemed, preferred to join the Chinese bandwagon. Neither choice allowed for independence of action.

North and South Korea were not part of the bandwagon, for once again they were engrossed by their mutual obsession with inter-Korean relations. This was the only thing they shared. North Korea had an economy moving in reverse while South Korea's forward economic growth was only slowing. North Koreans were starving, while South Koreans were free to watch the 'soap opera' trials of two of their previous Presidents. Behind their iron veil, North Korea's leaders seemed locked in a deepening factional struggle, while South Koreans were treated to a new range of scandals and the by now well-known game of 'musical ministers'.

The horizon of these two insular Koreas barely encompassed North-east Asia and its interested great powers. Both Koreas vied for the favours of the United States and, to a lesser extent, of China. Little progress was made in bringing Pyongyang and Seoul, plus the two great powers, into formal talks, but efforts towards *détente* were sustained for fear that domestic unrest, especially in North Korea, could bring the region to the brink of war.

The stakes were far less serious in South-east Asia. In Indonesia, there were increased worries about the risks of instability surrounding the succession to President Suharto. Several smaller ASEAN states suffered economic decelerations, and the swiftness with which these occurred was taken as a warning of just how quickly the glory days of growth can end. Unease in ASEAN about sustaining economic growth in quasi-authoritarian political systems made the Association blasé about contemplating the admission of new members from Indochina and of Myanmar where authoritarianism was far more ruthless. For ASEAN states, domestic particularism had long been manifest in the robust defence of a Victorian sense of state sovereignty that decried any hint of interference in the domestic affairs of another state.

The concentration on domestic affairs was also manifest in a tortoise-like approach to regionalism. The ASEAN Regional Forum (ARF) widened without deepening. In the wake of the Asia–Europe Meeting (ASEM) in 1996, and as part of the continuing process of regional ASEM consultations, for the first time the East Asians began caucusing without Caucasians. However, the result was an effective veto by China and others wanting to defend the virtues of state sovereignty and minimise the need to accept the constraints of international interdependence.

South Asia was no exception to the regional focus on internal affairs but, unfashionably, it offered more hope for linking domestic and international affairs positively. While Pakistan's outlook remained limited in scope, and thus promised only more of the same corruption and misrule, in India there was a sustained determination to reform and liberalise the economy in order to join in global interdependence. Indian economic reforms were still at an early and fragile stage, but in 1996 they were all the more remarkable for their ability to survive an inconclusive parliamentary election and a decentralised federal system. India exemplified the possibility that paying close attention to home needs – the urgency of reviving a flagging economy – was driving greater openness and interdependence. In New Delhi, democratic politicians sought legitimacy through prosperity and connections with the outside world; their success would be an important lesson for leaders throughout East Asia to contemplate.

Fraying Multilateralism

ASEAN's efforts to create regional security through consensus has run up against the concern of some East Asian states that dialogue and consensus is not enough to assure their security. Indonesia, once a leader of non-alignment, has become sufficiently concerned to agree a bilateral security treaty with Australia. Neither country says so, but the agreement represents some insurance against a rising China. Australia also has new military agreements with Singapore, and other East Asian nations are contemplating moving away from total reliance on multilateral arrangements like the ARF towards more traditional balance-of-power arrangements.

Other regional multilateral agreements are also fraying. The coalition that defeated Saddam Hussein and which has since held together in maintaining sanctions against the constant threat he represents is no longer robust. France and Russia have both argued for lifting the sanctions; Egypt and other Arab nations are now questioning the continuing usefulness of the coalition. And whether NATO or the EU can sustain the Dayton Accords in Bosnia after US troops withdraw next year is very uncertain.

The limits of multilateralism as a driving force are starkly illustrated by how unsteady these existing multilateral endeavours now are. Even more telling is the reluctance of the international community to undertake any new initiatives as violence erupts in economically backward countries. Even humanitarian efforts – supplying needed food, medicine and shelter – is being questioned as countries without functioning governments succumb to rule by heavily armed and uncontrolled mobs who threaten the lives of those trying to help.

In the face of the chaos that swept through Albania and Zaire in the first months of 1997, outside governments cited the usual litany of reasons for their lack of involvement. There was no useful government to identify or support, their own military forces were stretched by engagement in Bosnia and elsewhere, their defence budgets were inadequate to support further actions, and they feared becoming mired in a never-ending intervention. These reasons were no less true for being familiar. Yet they amounted to a recognition that there was no support at home for dangerous action abroad. Unless crises could be shown to have a direct effect on a state's own people or security, no government today felt impelled to band together with others to secure collapsing regimes.

A Not So Brave New World

It is as true as ever that if the United States does not lead, very little is done globally. For this reason, ubiquitous domestic political struggles ripple out from Washington and cause heavy waves elsewhere. Without strong, coherent and consistent leadership the US electorate is now content to give all its attention to its provincial concerns. But President Clinton's ability to lead is constrained both by the challenges to his influence and by the lack of a concrete security threat. The contemporary international scene may be unstable and call out for firm management, but the instability is amorphous and resistant to clear definition.

The task for statesmen, however, is to supply that clarity. If it can be found, action can be taken. NATO is certain in 1997 to invite new members to join its club, just as the EU is certain not to. It can be argued, with some merit, that NATO enlargement is not the proper response to the questions posed by world stability and security. But even if that is the case, what it does make clear is that vision and conviction can still be translated into action. Unless leaders raise their eyes above the quotidian tasks, however, the predilection for parochial behaviour will triumph.

Parochial behaviour can, of course, have extrovert implications. In 1996, the US Congress, through the Helms–Burton and D'Amato legislation, formally interested itself not only in the domestic politics of other states, but in constraining the way foreign countries – and their private citizens – dealt with so-called 'rogue' states. A parochial Congress could still be extra-territorial, and as such could strain alliance relationships, and even multilateral institutions, that have always been considered central to US well-being. A measure of US power today is that its impact on world events can be almost as great when it is looking inward as when it is looking outward. What the events of 1996 overwhelmingly showed was that no state, including the US, was immune from foreign meddling in domestic affairs or domestic meddling in foreign affairs. The distinction between foreign and domestic is indeed becoming thin and the net result is more incoherence.

Strategic Policy Issues

Assessing the Cruise Missile Threat

Although the spread of land-attack cruise missiles has attracted much less international attention than ballistic-missile proliferation, there is growing concern that cruise missiles may eventually eclipse their ballistic cousins – or, at the very least, share centre stage with them – as a major security challenge. How quickly cruise missiles may compete with ballistic missiles for the attentions of policy-makers is of more than just academic interest. The mere fact that ready sources of proliferation exist and that serious threats could emerge at the same time as first-generation cruise missiles suggests that considerable thought must be given to missile defence.

Energised by Iraq's *Scud* missile attacks during the 1991 Gulf War, the United States has invested billions of dollars in ballistic-missile defences; and it has urged its allies around the world to deploy such defences as well. By contrast, the US has spent less than 10% of the total invested in ballistic-missile defence on improving defences against land-attack cruise missiles. To determine how appropriate these relative priorities in missile defence may be, several questions must be addressed: what distinguishes the cruise-missile threat from ballistic-missile proliferation? What are the sources of cruise-missile proliferation and at what pace is the threat likely to unfold? Given huge investments in air defences against manned aircraft, why are additional resources needed to cope with land-attack cruise missiles? And, finally, just how amenable is cruise-missile proliferation to arms-control measures?

Differences Between Cruise and Ballistic Missiles

Understanding the differences between cruise and ballistic missiles helps explain why cruise-missile proliferation could become at least as severe a threat as ballistic missiles are already – and perhaps even more so. Cruise missiles are small unmanned aircraft. They may be powered by rockets or jet engines, while being guided and navigated by internal computer or remote control. All cruise missiles stay aloft by aerodynamic lift and are powered and guided all the way to their target. By contrast, ballistic missiles are powered by rockets that operate only during the initial boost phase. Once the rocket cuts off, the missile coasts to the target along a ballistic trajectory and – except for very sophisticated systems – can no longer be guided.

The public has now become aware of both ballistic and cruise missiles as a result of their prominence in recent combat: in the former case, dramatic television scenes of Iraqi *Scud* missiles dropping on Arab and Israeli targets during the 1991 Gulf War; and, in the latter case, US *Tomahawk* missiles used during and after *Operation Desert Storm* against Iraqi targets and then again to support NATO air-strikes in Bosnia in September 1995. But such uses are by no means new. Ballistic and cruise missiles were first used in combat over half a century ago. The Nazi V-2 ballistic missile took a decade to develop and was successfully launched about 3,000 times against the allies during the Second World War, causing over 9,000 casualties in London alone. By comparison, the earlier and much simpler V-1 cruise missile took only a year to develop at about one-hundredth the cost of the V-2. About 21,000 V-1s were successfully launched against the allies during the Second World War, causing more than 18,000 casualties in London – although nearly three-quarters were eventually shot down

Today, acquiring ballistic missiles is becoming more complex and costly than developing or buying cruise missiles, and – largely due to the success of export controls – will arguably become even more difficult as time passes. Since the Missile Technology Control Regime (MTCR) was established in 1987, the major industrial powers no longer provide the kind of systematic aid considered essential to complete indigenous ballistic-missile programmes in the developing world. These strictures do not apply in the same way to cruise missiles. Only the advanced industrial powers possess – and tightly control – the sophisticated terminal-guidance technologies needed to make ballistic missiles sufficiently accurate for use with conventional warheads. Without the widely proliferated *Scud* missile, essentially a 1950s Soviet improvement on the Nazi V-2, ballistic-missile proliferation would not be the problem it is today.

Although the range of the *Scud* and its many derivatives has been extended, its poor accuracy reduces its effectiveness in delivering conventional explosives. The *Scud*'s main advantage is its high-delivery velocity, which enables it to penetrate all but the latest theatre missile defences. Yet this high-delivery velocity, not to mention poor accuracy, makes it difficult for these ballistic missiles to adjust the pattern of chemical and biological agents they release according to wind directions and target shape. Thus, their overall value is limited largely to political terror, while the incentive for acquiring them probably has most to do with gaining prestige and possessing a weapon, however militarily ineffective, that can sustain operations relatively free from the counterforce efforts of an adversary – as Saddam Hussein's *Scuds* managed during the Gulf War.

Until recently, policy-makers and analysts have been right to focus on ballistic- rather than cruise-missile proliferation. Most aerodynamic missiles produced since the Nazi V-1 are used to attack ships and aircraft or to defend coastal areas. Over 70 countries, 40 of them in the developing world, possess over 75,000 anti-ship cruise missiles, most with ranges under 100km. Only a dozen countries have land-attack cruise missiles approaching the quality of the US *Tomahawk*. This is primarily because tight export controls have effectively prevented the spread of highly accurate guidance schemes like Terrain Contour Matching (TERCOM) systems or Digital Scene-Matching Area Correlation (DSMAC) systems.

However, a stark new reality is changing the missile-proliferation challenge: military breakthroughs are increasingly resulting from commercial rather than secret military research. Nowhere are the consequences of this more evident than in the unencumbered flow of the enabling technologies needed to develop highly accurate land-attack cruise missiles. Chief among these new commercial technologies are cheap guidance and navigation systems based on the US Global Positioning System (GPS), whose use during *Operation Desert Storm* gave a brief glimpse of the revolutionary changes now under way in warfare. Combined with commercially available mission-planning tools and one-metre-resolution satellite imagery to target fixed objects, new guidance and navigation technology for cruise missiles offers substantially more accurate delivery (by at least a factor of ten) and costs substantially less (by a half or more) than far more complex – and export-controlled – ballistic-missile guidance systems. Before the advent of these new technologies, developing states had little incentive to build or acquire inaccurate land-attack cruise missiles, particularly when the former Soviet Union was so generously dispensing *Scuds* and related production wherewithal to its client states.

In addition to militarily significant accuracy in delivery, land-attack cruise missiles offer other compelling operational advantages compared to existing and prospective ballistic missiles. Cruise missiles can be placed in canisters, making them fairly easy to maintain and operate in harsh environments. Their relatively compact size suggests more flexible launch options, more mobility for ground-launched versions and a smaller logistics burden, which could make them even less susceptible to counterforce targeting than were Iraqi *Scuds* during *Operation Desert Storm*. In addition, cruise missiles need not be stabilised at their launch points and can be fired from commercial ships and aircraft, as well as from ground launchers. Perhaps most important of all, the cruise missile's aerodynamic stability, which makes it an inherently easier and cheaper platform from which to deliver conventional payloads accurately, also makes it a better platform for effective dispersal of chemical and biological agents. The lethal area for a given quantity of biological agent delivered by a cruise missile can be at least ten times greater than that delivered by a ballistic missile.

Sources of Cruise-Missile Proliferation

The flow of critical enabling technologies is a necessary but not sufficient condition for the widespread proliferation of land-attack cruise missiles. The rate at which developing nations absorb and assimilate openly available technologies and, most importantly, integrate them into militarily significant weapon systems, are equally if not more relevant. But indigenous development from scratch is not the only source of cruise-missile proliferation. Developing nations could also convert existing anti-ship cruise missiles or unmanned aerial vehicles (UAVs), or directly purchase complete land-attack cruise missiles now being sold by key industrial world producers.

Indigenous Development

Indigenous development is clearly the longest route to a militarily significant arsenal of cruise missiles, but it is becoming easier as the civilian and military aircraft industry becomes increasingly globalised. The structures, propulsion, autopilots and navigation systems used in manned aircraft are essentially interchangeable with those of cruise missiles. In its 1991 study, *Finding Common Ground: US Export Controls in a Changed Global Environment*, the US National Academy of Sciences found there had been significant growth around the world in aircraft maintenance capability – a critical indicator of the spread of technical aircraft expertise. The Academy concluded that the globalisation of the aircraft industry had 'negative implications for control by any single nation of the export of production technology'. To make matters worse, developing nations are aggressively taking advantage of today's 'buyer's market' in aerospace to demand offsets with their purchases that will give them indigenous aircraft maintenance and even production capabilities. And if nations choose to forgo new aircraft purchases to upgrade existing craft, they can buy engines, advanced electronics and other key aerospace subsystems at bargain prices. Russia alone expects about $10 billion in annual sales of spare parts and maintenance services for Soviet-designed aircraft.

Converting Anti-ship Cruise Missiles

Although these trends will not allow developing nations to build land-attack cruise missiles from scratch in the short term (5–10 years), they will help to reduce the amount of time it might take to modify existing anti-ship cruise missiles or UAVs. Because they are so ubiquitous, anti-ship cruise missiles are the most direct conversion route. As noted, there are roughly 75,000 anti-ship cruise missiles in existence today, but only a proportion can be transformed into land-attack systems with ranges of over 300km. This is because modern designs, like the French *Exocet*, are densely packed with integrated electronics and software, leaving little room for adding fuel, changing engines or rearranging avionics. By contrast, anti-ship cruise missiles like the Russian *Styx* and its Chinese derivative, the *Silkworm* family, are inherently easier to modify and require far less manufacturing skill to do so because of their sheer size and simplicity of design. Structural modifications would require producing bulkheads and riveting simply shaped aluminium plates. In addition, replacing the original and bulky autopilot and avionics with a combined GPS and inertial guidance system could not only achieve terminal accuracies of less than 100m, but also free up additional space for extra fuel that would greatly extend the missile's range.

From the standpoint of weapons proliferation, the *Silkworm* cruise missile is similar to the *Scud* ballistic missile in that it is available globally, including to countries like Bangladesh, Dubai, Egypt, Iran, Iraq, North Korea, Pakistan and Zaire. China has established *Silkworm* production facilities in Egypt, Iran and North Korea. Iran, for its part, is reported to be working on extending the *Silkworm*'s range to 400km. In 1995, Tehran also announced that it had test fired an indigenously produced anti-ship cruise missile. North Korea currently exports the *Silkworm* and is working on more advanced cruise missiles. In June 1994, US officials stated that Pyongyang had tested a 160km-range anti-ship cruise missile, probably an improved version of the Chinese-furnished HY-2 or HY-4. During the late 1980s, Iraq had already built on its earlier acquisition of Russian *Styx* and Chinese *Silkworm* missiles to develop its own series of longer-range variants (approximately 200km) by lengthening the missile's body and adding fuel.

The most direct route to transforming an anti-ship cruise missile into a much longer-range land-attack system would be to use the Chinese turbojet-powered HY-4 *Silkworm* derivative, which China currently offers for sale on the export market. This missile can be ground- or air-launched and has a payload of 500kg and a range of 150–200km. Alternatively, the liquid-rocket HY-1 and HY-2 could be transformed, but to extend their range significantly (to about 500km or more) would require replacing their existing liquid-rocket engines with a suitable turbojet engine. If such engines could not be purchased directly from China, Canadian, European, Japanese, US and other international firms produce numerous unrestricted gas turbine engines for the civilian and military markets that would be suitable replacements.

Overall, however, if increased range is the desired objective, developing nations are likely to turn to *Styx* or *Silkworm*-class missiles already in their inventories, or newer Chinese HY-4s, in an attempt to lessen the burden of acquisition. Otherwise, modifications of smaller, more modern anti-ship cruise missiles are only likely to achieve land-attack ranges of 200km or so – which is still significant in confined geographic areas like the Middle East or the Persian Gulf.

Figure 1 *The Missile Technology Control Regime*

- The Missile Technology Control Regime (MTCR) was established in 1987 by the US and its Group of Seven partners (Canada, France, former West Germany, Italy, Japan and the United Kingdom). There are currently 28 members of the MTCR and three 'adherents' – China, Israel and Ukraine

- The MTCR seeks to limit the proliferation of missiles, unmanned air vehicles (UAVs) and related technologies capable of carrying a payload of 500kg for at least 300km (this requirement is even stricter with regard to states with potential weapon-of-mass destruction capabilities)

- The regime was extended in 1993 to cover all systems capable of delivering nuclear, chemical and biological weapons

- As a regime, the MTCR does not entail formal treaty obligations, but consists of a common export policy applied to two categories of items relevant to missile development, production and operation

- Category I includes complete rocket systems, UAVs (including cruise missiles, target and reconnaissance drones) and 'complete subsystems' such as rocket engines and re-entry vehicles

- Category II includes parts, components and subsystems, including propellants, test equipment and flight instruments

- The commitment to restrict the export of these items is implemented nationally within the context of participating states' export laws

Direct Purchase from Major Industrial Suppliers

Between them, China, France, Italy, Russia, the United Kingdom and the United States have exported anti-ship cruise missiles and UAVs to more than 40 developing nations. While such transfers are useful stepping-stones to land-attack cruise missiles and are generally not subject to any MTCR controls, the quickest way for developing nations to obtain land-attack cruise missiles is to purchase them directly from an industrial-world supplier. A major shortcoming in the MTCR's handling of trade-offs between range and payload for cruise missiles threatens to exacerbate the spread of land-attack cruise missiles with very advanced features.

The MTCR is most restrictive in its treatment of missiles capable of carrying 500kg payloads to ranges of 300km or more. In this regard, MTCR member-states are obliged to make a 'strong presumption to deny' such transfers. But because the threshold was designed with ballistic missiles in mind, it is much better suited to dealing with trade-offs between ground-to-ground ballistic missiles than air-launched cruise missiles. For example, variations in cruise-missile flight profiles can lead to quite different ranges. Moreover, from an engineering standpoint, it is relatively easier to 'scale-up' the range of an existing cruise-missile system than that of a ballistic missile. The technology required to produce a 1,000km-range cruise missile is not fundamentally different from that needed for very short-range systems. Transforming short-range *Silkworm* anti-ship cruise missiles into much longer-range land-attack systems illustrates this point. Given the modularity and inherent flexibility of cruise missiles, several more variables contribute to trade-offs between range and payload than is the case for ballistic missiles.

Possible confusion over the extent to which MTCR provisions apply to air-launched cruise missiles may account for France's apparent willingness to consider exporting its *Apache* land-attack cruise missile – although no known transfers have yet occurred. Under development since 1989, the *Apache* is modular in design and will be produced in several variants, all of which use the same airframe. The short-range variant, the apparent candidate for export, has a nominal range of 140km with a payload of 520kg. The national variant, earmarked for French use only, drops its payload to around 400kg and increases its range to 600km. Another version, *Storm Shadow*, won the UK's Conventionally Armed Stand-Off Missile (CASOM) competition in 1996, and will be developed and produced jointly with British Aerospace. Range variations aside, the *Apache* has a stealthy aerodynamic shape, low infra-red signatures and a combination of guidance and navigation schemes designed to achieve high-terminal effectiveness. Should such a system fall into the wrong hands, not only would it provide design insight into an advanced-technology missile, but also a robust threat system capable of challenging the most advanced air-defence systems.

Given that France is a charter member of the MTCR, its handling of *Apache* exports is likely to shape the behaviour of other exporting states which are either developing or considering exporting land-attack cruise missiles. At the Abu Dhabi Defence Exhibition in 1993, Russia offered for sale a shorter-range version of the 3,000km-range AS-15 cruise missile. In an apparent effort to prevent the missile falling under the MTCR's most restrictive provisions, Russia advertised a payload of 410kg and a missile range just exceeding 500km. In addition, the variant was purported to include several advanced features, foremost a TERCOM-like guidance system combined with position updates from Russia's GPS-equivalent Global Navigation Satellite System, GLONASS, which promised a terminal accuracy of less than 20m.

Table 1 *Selected Aerodynamic Missile Arsenals and Programmes*

Country	Missile	Range (km) / Payload (kg)	Type	Guidance	Propulsion	Status
China	HY-4 *Sadsack*	150 / 500	ASCM	I, AR	Turbojet	In service
	XW-41	300* / 500*	ASCM	GPS*	Turbojet	Proposal
	Delilah 2	385* / 450*	LACM	I, GPS, IIR*	Turbojet*	Under development with Israel
	C-802 land-attack*	300* / *	LACM	GPS, TM*	Turbojet	Under development*
France	*Apache*	140–600/400–520	LACM	I, GPS, TM, IIR	Turbojet	Under development
India	SSC-3 *Styx*	80 / 500	ASCM	I, AR or IR	Rocket	In service
	Lakshya	500 / 200	TD	C, GPS	Turbojet	Production
	Lakshya 2	600 / 450	LACM	GPS*	Turbojet	Under development
Iran	C-802 *Saccade*	120 / 165	ASCM	I, AR	Turbojet	IOC 1996*
	HY-2 *Seersucker*	95 / 500	ASCM	I, AR or IR	Rocket	In service
	SS-N-22 *Sunburn*	110 / 500	ASCM	I, AR	Rocket	In service
	Silkworm upgrade	400* / 500*	LACM*	GPS*	Turbojet*	Under development
Iraq	HY-2 *Seersucker*	95 / 500	ASCM	I, AR or IR	Rocket	In service
	Otomat Mk1	60 / 210	ASCM	AR	Turbojet	In service
	FAW-70	80 / 500	ASCM	I, AR	Rocket	IOC 1990
	Ababil	500* / 200*	LACM	TV*	Turbojet	IOC 1988*
Israel	*Gabriel* 3	35–60 / 150	ASCM	I, C, AR	Rocket	In service
	Gabriel 4	200 / 150–200	ASCM	I, C, AR	Turbojet	Under development
	Popeye 1	85 / 360	LACM	I, TV or IIR	Rocket	In service
	Popeye 3	350 / 360*	LACM	I, TV, or IIR	Turbojet	Under development

Table 1 *continued*

Country	Missile	Range (km) / Payload (kg)	Type	Guidance	Propulsion	Status
Libya	SSC-3 *Styx*	80 / 500	ASCM	I, AR or IR	Rocket	In service
	Otomat Mk2	180 / 210	ASCM	I, C, AR	Turbojet	In service
	AM-39 *Exocet*	50 / 165	ASCM	I, AR	Rocket	In service
North Korea	HY-2 *Seersucker*	95 / 500	ASCM	I, AR or IR	Rocket	In service
	C-801 *Sardine*	40 / 165	ASCM	I, AR	Rocket	In service
	HY-2 upgrade	160 / 500	ASCM	*	Rocket*	Under development
Russia	Kh-55SM	3,000 / Nuclear (200 KT)	LACM	I, TM	Turbofan	In service
	Kh-65SE	500–600 / 410	LACM	I, GLONASS, TM	Turbofan*	Proposal
	AFM-L *Alpha*	250* / 200*	LACM	I, GLONASS*, AR	Turbojet*	Under development*
Serbia	SS-N-3B *Sepal*	450 / 1,000	ASCM	C, AR or IR	Turbojet	In service
	RBS-15	90 / 200	ASCM	I, AR	Turbojet	In service
South Africa	*Skua*	800 / 100	TD	I, C, GPS	Turbojet	In service
	MUPSOW	150* / 400*	LACM	GPS*	Turbojet	Under development
Syria	SSC-1 *Sepal*	450 / 1,000	ASCM	C, AR or IR	Turbojet	In service
	SSC-3 *Styx*	80 / 500	ASCM	I, AR or IR	Rocket	In service
Taiwan	*Hsiung Feng* 2	80 / 225	ASCM	I, AR, IR	Turbojet	In service
	Hsiung Feng 3	200 / *	ASCM	*	Turbojet	Proposal

United Kingdom	*Centaur*	* / 225	LACM	I, GPS, IIR	Turbojet	Under development
	Tomahawk, TLAM-C, Block III	1,650 / 320	LACM	I, TM, GPS, DSMAC	Turbofan	In service
United States	AGM-86B ALCM	2,500 / * Nuclear (200 KT)	LACM	I, TM	Turbofan	In service
	AGM-86C CALCM	2,000 / 450	LACM	I, GPS	Turbofan	In service
	Tomahawk TLAM-N	2,500 / * Nuclear (200 KT)	LACM	I, TM	Turbofan	In service
	Tomahawk TLAM-C, Block III	1,650 / 320	LACM	I, TM, GPS, DSMAC	Turbofan	In service
	Tomahawk TLAM-D, Block III	1,300 / * Submunitions (250kg)	LACM	I, TM, GPS, DSMAC	Turbofan	In service
	Tomahawk TASM	480 / 450	ASCM	I, AR	Turbofan	In service

Notes:

AR Active radar
ASCM Anti-ship cruise missile
C Command
DSMAC Digital Scene-Matching Area Correlation
GLONASS Global Navigation Satellite System
GPS Global Positioning System
I Inertial
IR Infra-red
IIR Imaging infra-red operating capability
IOC Initial
KT Kiloton
LACM Land-attack cruise missile
TD Target drone
TM Terrain matching
TV Television

* An asterisk denotes probable data where accurate information is not known

Sources: Duncan Lennox (ed.), *Jane's Strategic Weapon Systems* (Coulsdon, Surrey: Jane's Information Group, 1997); Robert Shuey, 'Ballistic and Cruise Missile Forces of Foreign Countries', *Congressional Research Service Report for Congress*, 95-688 (Washington DC: US Library of Congress, 25 October 1996); Robin Ranger, *et al.*, *Cruise Missiles: Precision and Countermeasures* (Lancaster: Lancaster University, Centre for Defence and International Security Studies, 1995); and General Accounting Office, *Cruise Missiles: Proven Capability Should Affect Aircraft and Force Structure Requirements*, GAO/NSIAD-95-116 (Washington DC: US General Accounting Office, April 1995).

Israel – an MTCR adherent – is reportedly transforming, with rumoured Chinese financial assistance, its *Delilah* UAV into a 400km-range air-launched cruise missile. The Spanish company CASA has announced its intention to build a land-attack cruise missile with characteristics strikingly similar to the French *Apache*. By relying almost exclusively on commercial off-the-shelf technologies, CASA hopes to keep unit costs low enough to compete favourably with the *Apache* on the export market. Until greater consensus is gained amongst MTCR member-states over how to deal with cruise-missile exports, the direct purchase of advanced land-attack cruise missiles from large industrial states may become the major source of cruise-missile proliferation.

Monitoring the Pace of Cruise-Missile Proliferation

However numerous the sources of cruise-missile proliferation, land-attack cruise missiles have not yet emerged as a significant threat. Decisions to acquire improved air defences against cruise missiles traditionally rest on the intelligence community validating (or formally approving) the threat. Such validation signifies that the threat has emerged both in sufficient numbers and sufficient quality to warrant expenditure to counter it. There is growing concern, however, that cruise missiles could emerge quickly, with little or no advanced warning, to threaten existing defences, as Iraq's modified *Scud* ballistic missiles did – politically if not militarily – in the Gulf War.

Interestingly, deciding where the threat is coming from – triggered by a US National Intelligence Estimate (NIE) – dominated the missile-defence debate in America during 1996. Although NIE 95-19, 'Emerging Threats to North America During the Next 15 Years' published in November 1995, dealt primarily with ballistic-missile threats, its discussion of how such threats unfold illustrated just why cruise-missile developments could surprise decision-makers.

Developing a ballistic missile involves discrete sequential steps that become increasingly demanding in terms of the skill and time required for missiles over 1,000km range. Because testing is so critical for ballistic missiles generally, and long-range systems especially, and so easy to detect, NIE 95-19 concluded that the intelligence community could give decision-makers at least five years' warning of the deployment of a new inter-continental-range ballistic missile (ICBM).

Comparable warning time would very likely not be available in monitoring cruise-missile developments. NIE 95-19 noted that a potential proliferator could use cruise missiles earmarked for regional war-fighting and adapt them for launch from ships capable of threatening US territory. Such cruise-missile threats, the NIE pointed out, would be easier and less detectable than developing ICBMs because there is no need to test large rocket motors. Whether or not a proliferating state would choose to threaten

North America in this way, land-attack cruise missiles remain comparatively easier to build and hide than ballistic missiles because the MTCR intentionally avoids restricting manned aircraft sales and production. This means that developing countries can readily intermingle cruise-missile developments with legitimate aircraft purchases or production, making detection problematic at best.

Difficulty in its detection is not the only reason why cruise-missile development will be difficult to monitor. While ballistic-missile improvements tend to unfold sequentially over time, several levels of cruise-missile capability – including both early missile designs and advanced missiles with stealth features – could emerge virtually simultaneously. Indeed, because the MTCR does not restrict sales of advanced cruise missiles, such systems could appear in developing-world arsenals even before indigenously produced or transformed anti-ship cruise missiles emerge within the next decade.

Defending Against Land-Attack Cruise Missiles

The United States and its allies have invested huge amounts in air defences, including ground and airborne surveillance, fighter-based air-to-air missiles (AAM), and surface-to-air missiles (SAM). However effective these investments may be in coping with manned aircraft threats, there are definite shortcomings today in defending against low-flying cruise missiles. These defects will become even more exaggerated as low-observable capabilities (measures to reduce radar and infra-red signatures) become a more prominent feature of new cruise missiles.

To appreciate the nature of the cruise-missile threat, consider how Iraq might have used land-attack cruise missiles to create far greater havoc for the Western air defences than was the case during the Gulf War. Because Iraq's air threat during *Operation Desert Storm* was essentially non-existent, allied defences could focus on high-angle *Scud* ballistic-missile trajectories, which are easy for radars to distinguish from aircraft threats. But had Iraq used both ballistic and cruise missiles, coalition air and missile defences would have been severely tested by the need to identify friendly aircraft from similarly low-flying enemy cruise missiles. The inadvertent shooting down by friendly aircraft of two US Army *Blackhawk* helicopters over northern Iraq in 1995 attests to the continuing problem of positive combat identification and consequent friendly-fire casualties. In short, introducing land-attack cruise missiles adds a complicated new dimension to air battles, no matter how large or small the enemy cruise missile's radar signature.

Ideally, the best way to defend against cruise missiles is to detect them as early as possible after their launch. Counterforce operations against Iraqi *Scuds* proved almost useless, except in reducing the *Scud* launch rate. Countering ground-launched cruise missiles before their launch will be even more challenging. Cruise missiles will not provide a detectable launch signature like ballistic missiles do, nor will they be constrained to operate from pre-surveyed sites. Thus, in-depth defence will be needed to engage cruise missiles as close as possible to their launch points. If incoming cruise missiles can be positively identified or tracked early on, a layered defence can then engage them either immediately after their launch, mid-way during their flight or in the terminal phase. Effectiveness in each phase contributes to a low leakage rate, which is especially critical if cruise missiles are armed with nuclear, biological or chemical warheads.

Current Cruise-Missile-Defence Shortcomings

There are serious shortcomings in the ability of current theatre air defences to provide such a layered defence. Such air defences consist of fixed-wing airborne surveillance aircraft (such as airborne warning and control system aircraft (AWACS) and the Navy E-2C), fixed-wing fighter aircraft armed with AAMs, land- and sea-based SAMs and associated surveillance and fire-control radars, and various systems for battle management and command, control and communications. A key weakness is that airborne surveillance assets are not internetted with SAMs, meaning that the picture ground-based SAMs like *Patriot* see is not a broad one furnished by the elevated airborne platform (hundreds of kilometres), but one hindered by the line-sight range to the horizon of *Patriot*'s associated radar (25km or less). The same kind of constraint hinders fighters from engaging cruise missiles at long ranges. Airborne surveillance cannot currently cue fighters or furnish fire-control information to fighter-launched AAMs. Thus, SAMs and AAMs can only engage cruise missiles at relatively short range, which obviates any notion of an in-depth layered defence.

Current air-defence limitations create enormous opportunity costs. Unacceptably large numbers of SAMs, AAMs and fighter aircraft, needed for other missions besides cruise-missile defence, would be required even to approach creating a meaningful, wide-area defence against cruise missiles. Moreover, while current air-defence systems may have some capability against first-generation cruise-missile threats, problems of combat identification, left uncorrected, may well cause friendly-fire casualties. Worst of all, as cruise missiles with stealthy features enter adversary arsenals, not only will improved combat identification and system internetting be needed, but upgraded detection means will be required as well.

Confronting Hard Choices

Both the US Department of Defense (DoD) and Congress found common cause in 1995 in recognising the need to improve defences against the emerging threat of land-attack cruise missiles. In January 1995, the DoD issued a *Report of the Defense Science Board Summer Study Task Force on Cruise Missile Defense* outlining many of the weaknesses discussed above. Congress, for its part, fashioned a 'Cruise Missile Defense Initiative' in its

National Defense Authorization Act for Fiscal Year 1996, calling on the Secretary of Defense to develop a coordinated approach to deploying affordable and effective defences against both ballistic- and cruise-missile threats.

Extraordinarily tight budgets and pressure from military services to modernise expensive platforms greatly complicate decisions about cruise-missile defence priorities. And because there is uncertainty over just how quickly the cruise-missile threat will emerge, there is a corresponding reluctance to reprogramme funds until there is greater clarity about the threat. Finally, truly affordable and effective cruise-missile defences, fully integrated with defences against ballistic missiles, will foster time-consuming bureaucratic jousting over service roles and missions and new joint doctrine for a fully netted air and missile-defence system. Indeed, some of the new technologies central to improving cruise-missile defences suggest entirely new operational concepts that threaten rather than excite conservative service organisations.

With these frictions in mind, defence decision-makers face a range of choices for improving existing cruise-missile defences. The best way to extend battlespace for a layered defence is to improve the performance of existing airborne sensors against low-flying cruise missiles. To enhance affordability and operational flexibility, it probably makes sense to consider a mixture of more expensive fixed-wing sensors like AWACS and cheaper aerostat platforms (blimp-like balloons that use lighter-than-air gas for buoyancy, which can carry sensors). These have played an important role in detecting illegal drug activity along the United State's southern border. Although aerostats with improved sensors would be weather-limited and take time to deploy in far-off regional contingencies, they can complement faster-reacting, weather-insensitive, more mobile fixed-wing assets. Especially appealing is the fact that aerostats could furnish extremely long on-station times during lengthy pre-hostilities, or during low-intensity and peacekeeping operations, for a fraction of the cost of fixed-wing aircraft.

Effective wide-area defence will not be possible without netting together better elevated sensors and ground-based SAMs and fighter-based AAMs, each with improved seekers. This would permit these weapons to receive not only much better surveillance information, but high-quality fire-control data as well. Centralised fire control is perhaps the most revolutionary new concept for cruise-missile defence. Known as ADSAM (air-directed surface-to-air missile), an airborne fire-control sensor directs a ground-based SAM towards an aim-point, enabling the missile to intercept targets to its full potential range (100–150km), not just the 25km associated with its horizon-limited ground-based radar. The airborne fire-control sensor could also furnish the SAM with mid-course or terminal guidance updates, or the SAM could guide itself with its own on-board seeker. Using ADSAM, a single SAM battery would become a true area rather than a point-defence system by virtue of its ability – depending on the particular system – to provide defence for 10,000–70,000 km^2 of territory. Such a capability suggests that national defences against cruise missiles could become operationally and economically feasible for many US allies.

The ADSAM system necessitates adding new sensor technology for centralised fire control to either fixed-wing aircraft or aerostats. Besides fire control for air-directed SAMs, such an elevated platform could also furnish precision cues to fighter weapons to increase the effectiveness of their AAMs to their full potential range (roughly 60km). Each of the service's battle-management command-and-control data links would have to be closely integrated with such an enhanced system. The ADSAM concept is more than just a paper study; it was demonstrated in a US field test in early 1996. But to acquire a mix of advanced airborne sensors, air-directed SAMs and AAMs, together with greatly improved battle-management to produce a common air picture, will probably require an investment of several billion dollars, not to mention unprecedented cooperation within the Pentagon to forge a more effective joint doctrine on air defence for the three services.

Striking a Better Balance

Hard investment choices compounded by significant organisational conflict ensure that it could take up to a decade for the United States to make modest improvements in effective defences against land-attack cruise missiles. Complicating matters further is the extent to which US allies in Europe and Asia might want to render their own extended air-defence choices inter-operable with US systems. US allies are no less affected by these threat developments. A 700km-range land-attack cruise missile, launched from North Korea, could strike critical targets in Japan. A ship-launched version, fired from outside territorial waters, could strike most of the population and industry in Europe and North America. The non-governmental Gates Panel – formed at the direction of the US Congress to review NIE 95-19 and chaired by former CIA Director Robert Gates – drew attention to this threat when it presented its findings to the Senate on 4 December 1996. While the bipartisan Panel believed that the intelligence community made a strong case for why the US is unlikely to face an ICBM threat from a developing nation before 2010, it also concluded that not nearly enough attention was devoted to the currently feasible possibility that land-attack cruise missiles could be launched from ships within several hundred kilometres of US territory. This admonition applies equally to Europe and Asia.

No matter how quickly and to what extent the land-attack cruise-missile threat unfolds, policy-makers need to strike a better balance between ballistic- and cruise-missile issues. This applies as much to missile-defence investments as to developing a stronger consensus on arms-control prescriptions.

Even though uncertainty prevails over how quickly cruise-missile threats might emerge, the mere fact that the sources of proliferation are several and that advanced threats could emerge together with – or even before – first-generation cruise missiles suggests that prudent defence readiness steps need to be taken soon. Even full funding for robust theatre-wide cruise-missile defences will not begin to compare with costs for theatre-wide ballistic-missile defences. But it is a mistake to compare the two missions as if they were a zero-sum game. More effort is needed to understand the overlapping and complementary nature of technologies for cruise- and ballistic-missile defences to assure cost-effective choices. Finally, US allies need to be drawn into discussions about cruise-missile defence choices. Plausible cruise-missile threats compel these allies to think about inter-operability in coalition warfare and even national missile defences, which may prove more affordable against cruise than ballistic missiles.

Striking a better balance in the area of missile controls is also needed. Indeed, while it would be short-sighted, even dangerous, to depend too much on the MTCR to control cruise-missile proliferation, it is equally dangerous to let its current shortcomings *vis-à-vis* cruise missiles go unattended. A much stronger consensus for restricting ballistic rather than cruise missiles exists within the MTCR membership. Not only does this balance need to be rectified, but MTCR members urgently need to improve the regime's provisions in two specific ways. The first relates to clarifying how trade-offs are calculated between cruise-missile range and payload. Unless current ambiguities are reduced, advanced cruise missiles are likely to be exported without the same restrictions now in place for ballistic-missile transfers. The second improvement deals with tightening controls on stealth technologies for cruise missiles. The absence of strong controls in this area will make the technical challenges and costs of cruise-missile defence ever more demanding. Slowing the spread of these critical enabling technologies for highly effective cruise missiles will help defensive systems keep pace with threat improvements.

Biological Weapons: New Threats or Old News?

After decades of neglecting the potential threat of attacks by regional aggressors armed with biological weapons (BW), the discovery of Iraq's arsenal following the 1991 Gulf War reminded any doubters of the viability of this form of warfare. Six years after the Gulf War cease-fire, debate continues over how to assess the threat. Recent events, and the prospect of terrorist or large-scale BW attacks in future conflicts, have reinforced international concern.

In 1992, Russia admitted possessing a massive biological-weapons programme despite its commitment, as a member of the 1974 Biological Weapons Convention (BWC), to end all such activities. In Japan, before the now-infamous 1995 chemical-weapon attack on the Tokyo subway, members of the *Aum Shinrikyo* religious sect had reportedly experimented with biological agents by releasing small quantities of the lethal biological agent anthrax in Tokyo with no obvious lethal effect. In 1995–96 it became apparent that, before the Gulf War, Iraq had produced and converted for use in weapons enough biological agent to conduct military operations within the region. It was also feared that increasing instability in North Korea in 1996 and early 1997 might cause it to lash out, perhaps using biological or other weapons of mass destruction, in an attempt to destabilise South Korea and its allies.

A fundamental question has arisen: is there anything really 'new' about this emerging threat? If there is, it undermines the adequacy of traditional responses such as arms control and deterrence. To counter this emerging threat will thus require new approaches.

Figure 1 *Biological-Weapon Agents*

Agent Type **Causative agent (disease)**	**Symptoms/effects**	**Mortality (if untreated)**	**Onset**
Bacteria			
Bacillus anthracis (anthrax)	Cough, difficulty breathing, exhaustion, toxaemia, cyanosis, terminal shock	95–100%	2–4 days
Francisella tularensis (tularaemia)	Muscle ache, chills, cough, acute respiratory distress, exhaustion, prostration	30–40%	2–4 days
Coxiella burnetti (Q-fever)	Fever, pains, headache	0–1%	15–18 days
Virus			
Venezuelan Equine Encephalitis (VEE encephalomyelitis)	Joint pain, chills, headache, nausea, vomiting with diarrhoea, sore throat	0–2%	2–5 days
Toxin			
Botuinium Toxin (botulism)	Nausea, weakness, vomiting, respiratory paralysis	60–90%	0.5–3 days

Biological Agents and their Production

One fundamental concern in bacteriological warfare is how easily the basic material can be produced. A biological weapon typically uses micro-organisms (i.e., bacteria or viruses) or toxins. Bacteria are single-celled organisms that invade the human body, replicate and cause disease and death. Viruses do the same, but are much smaller, consist of genetic material (DNA or RNA) and must invade a host cell to multiply. Toxins are poisonous substances that can be synthetically produced or derived from living organisms. Weight-for-weight, BW agents can be hundreds to thousands of times more potent than chemical agents and can cause a variety of symptoms.

Knowledge of microbiology and the fermentation process are among the specialised skills required for BW agent production, but such skills are now becoming widely available in the medical and scientific industries and new fermentation techniques are well described in open literature. Moreover, biological-warfare agents can be cultivated by processes similar to those used for commercial biological products. Seed cultures for producing bacteria can be purchased from commercial vendors (Iraq, for example, purchased cultures from the United States) or extracted from natural sources, including animals. Most of the equipment used in cultivating BW agents is also commercially available for producing beer, food products, pharmaceuticals, vaccines, biopesticides and other similar products. Such equipment includes computer-controlled fermenters, centrifugal separators and freeze- and spray-dryers. Biological agents can also be grown using laboratory glassware, as was the case in Iraq.

Established biological and toxin agents can be mass-produced by almost any nation with a modestly sophisticated pharmaceutical or fermentation industry. Terrorist groups and individuals have already demonstrated their capability to develop such agents on a small scale. The spread of this capability has been facilitated in part by the revolution in biotechnology.

During the earlier part of the Cold War, Russia – and the United States until it gave up its efforts in 1969 – used expensive and labour-intensive techniques to produce biological weapons. Modern methods using more advanced technologies including, among others, continuous-flow fermenters with computer controls, real-time sensors and feedback loops, have vastly improved the productivity of firms manufacturing bacteriological products while also reducing the size of both the facilities required and the necessary capital investment. Similarly, recombinant-DNA techniques have made possible the production of militarily significant quantities of certain lethal toxins.

Knowledge of how to manipulate new techniques to mass-produce classical biological-warfare agents such as anthrax and botulinum has now spread well beyond the industrialised states. These countries, however, remain at the vanguard of the biotechnology revolution with their continuing efforts to manipulate genes and alter the genetic composition of cells. Their attempts to increase understanding of the structures and functions of the complex organic molecules that compose the human body will clearly advance medical science, and will inevitably and quickly spread as a result. The emerging knowledge base will be a boon, but it can also be abused.

Scientists will better understand how genetically modified pathogens and toxins attack the human body at the molecular level. The lethality of such agents could then be purposely increased; bacteria, viruses and toxins might also be modified for ease of storage, to produce more predictable battlefield effects, to enhance their survival in a variety of environmental conditions, and render them more resistant to treatment. And while the industrialised countries lead the biotechnology revolution, they no longer monopolise its potentially lethal by-products.

Biological Agent 'Weaponisation' and Delivery

The ability to produce microbial pathogens and toxins is within the reach of many countries, but converting an agent for use in a weapon is a far more challenging task. 'Weaponising' agents requires an intensive study of their military potential including infectivity, dosage and ability to survive in the atmosphere. It also requires converting most agents into a form suitable for delivery by a munition or other means.

Biological agents typically do not penetrate the skin and must be consumed or inhaled to infect or intoxicate a target population. Scare-mongers have promoted the idea of terrorists contaminating a municipal water supply, but this is largely a mythical threat. Reservoirs are purified with chemicals that would destroy all but the hardiest BW agents, and the dilution factor alone would render most established agents ineffective.

The most efficient approach to BW delivery is to disseminate the lethal agent in an aerosol form with droplets or particles that remain suspended in the air until reaching their target, and remain long enough in the pulmonary system after inhalation to be effective. Achieving the optimum size and weight for some agents can be a difficult technical challenge. Weapons can be constructed to deliver the agent in either a liquid or dry form; each has pros and cons. While liquid agent is easiest to produce, it is more difficult to disseminate. It can clog specially designed sprayers, does not remain viable in an aerosol for as long as dry agent, and thus will not cover as much territory.

Dry powders of pathogens or toxins are easiest to disseminate, occupy less volume in munitions and can be stored ready for use for longer periods than liquid agents. However, producing dry agent is more complex. It requires freeze-, spray- or some other type of drying, and the resulting cake must be milled to particles of the proper size. Dry agent can become airborne with ease and is therefore dangerous to handle during production. Without state sponsorship or access to the resources of a major commercial operation, the complexities and hazards of producing dry agent probably rule them out for most terrorist groups.

BW aerosol delivery means can range from simple liquid aerosol sprayers, either hand-held or mounted on a vehicle to manned aircraft, ballistic and cruise missiles. Manned aircraft and cruise missiles can use bombs for 'point-source' delivery or sprayers for 'line-source' delivery. A unitary bomb carrying liquid or dry agent can be fused to detonate either in the air or on impact. Hence, the agent is delivered to a specific point and then depends on the wind to carry it to its target. A more efficient method of point-source delivery is to use aerial bombs with a submunitions dispenser. 'Bomblets' of this type would typically be fused to detonate on impact and in a wide pattern, ensuring maximum dispersal of the agent.

The major disadvantage of bomb delivery is that impact alone is usually insufficient to ensure proper agent dispersal. Instead, some form of energy source is required to enhance dispersal. Explosive energy can be used, but this is inefficient as it can destroy a large percentage of the organisms or toxins carried. For instance, US bomb-makers found that some of their simple weapons achieved aerosol recovery rates (i.e., the percentage of agent that is disseminated in the optimum particle or droplet size for human inhalation) as low as 1–2%. Their inefficiency can be compensated by delivering more bombs or by using a non-explosive means to disperse the agents. One example is to disperse the agent from scattered submunitions by means of a gas. In the 1960s, the US developed a bomblet that on landing used a gas to expel liquid agent through a narrow opening.

Line-source delivery is the most effective means of covering large areas. Wing- or belly-tank spray dispensers, boom sprayers or slipstream dispersal from manned aircraft and cruise missiles can be used. Such devices enable agent to be dispensed in a long line, upwind of and tailored to fit the target. Tank dispensers and sprayers can deliver dry or liquid agent. For dry agent, a cruise missile or aircraft could use a simple slipstream dispersal method. The agent would be carried in a canister mounted on or inside the airframe; an intake would be opened at the forward end and the agent blown out from a port in the rear. For anthrax, the slipstream method has demonstrated a very high aerosol recovery rate.

Ballistic-missile delivery is clearly the most difficult of the three aerial delivery options. Ballistic missiles could deliver unitary warheads of bulk agent but, as noted, this method is highly inefficient and gives limited coverage for such an expensive option. Submunitions could also be delivered, but their design would be more complex than the type delivered by aircraft. To ensure an optimum pattern, submunitions with aerodynamic shapes or surfaces would have to be employed. This would enable them to disperse; otherwise they would follow the missile's ballistic trajectory and land in a less effective, closely spaced pattern. The pattern would depend further on fusing the missile to dispense the munitions at an optimum altitude. The submunitions themselves would have to protect the fragile micro-organisms or toxins from the heat of re-entry. A relatively sophisticated attacker, however, might overcome these obstacles. Cruise missiles are a better delivery means, given their low cost compared with manned aircraft and ballistic missiles, and their higher defence penetration capability compared to manned aircraft.

Once a would-be attacker has chosen his delivery method, meteorological conditions will strongly influence his attack planning. Micro-organisms are sensitive to sunlight, heat and humidity. Daytime attacks with BW thus risk a higher rate of agent decay, which would have an obvious effect on target coverage. Moreover, warmth creates rising vertical air currents that can carry particles and droplets up and away from the target area.

For these reasons, BW attacks are likely to achieve the greatest coverage per unit of agent if carried out at dusk or at night with a steady breeze to carry the agent over the target area. If BW munitions are air burst, this must occur below the inversion layer of air over the target; this stable layer is essential for keeping the pathogens and toxins at low altitude where they can be inhaled. To estimate the effects of his strike, the attacker must also calculate: the concentration of pathogens or toxins in his weapon; the aerosol recovery rate of his munitions or sprayer; wind speed; and the number of organisms or toxins required to kill or incapacitate individuals in the target area.

Figure 2 *Potential Effects of a Biological-Weapon Attack*

Agent	Purpose	Downwind reach (km)	Dead	Incapacitated
Coxiella burnetti (Q-fever)	incapacitant	>20	150	125,000
Francisella tularensis (tularaemia)	incapacitant	>20	30,000	125,000
Bacillus anthracis (anthrax)	lethal agent	>20	95,000	125,000

Japan's *Aum Shinrikyo* terrorists clearly miscalculated – their experiments with anthrax in Tokyo are not reported to have affected anyone; but a 1970 study by the World Health Organisation gives some indication of the possible effects of a biological-weapon attack. The scenario assumes a strike on a population centre of 500,000 people. Fifty kilograms of agent are delivered in a line source, upwind of the target and under optimal weather conditions. The results are outlined in Figure 2.

The Military Utility of Biological Weapons

Given the challenges associated with planning and executing BW attacks, the military utility of biological weapons has long been debated by experts. The strategic applications of BW seem less debatable than tactical ones.

The cost of fielding a large, modern conventional force is high and increasing. Nuclear weapons are very expensive and relatively difficult to develop, even for determined proliferators. Given the amount of chemical agent required to acquire a militarily significant stockpile (i.e., measured in tons), chemical weapons are far more expensive and pose greater logistical problems than biological weapons, which require tens of kilograms to carry out large-scale attacks. Hence, a BW arsenal could be seen as a cost-effective strategic deterrent against major powers, or against regional adversaries armed with other weapons of mass destruction. BW are particularly effective as a threat to unprotected civilian populations.

Beyond deterrence, biological weapons might also be useful for attacking large targets of strategic value during or prior to hostilities. Prior to a campaign, troops engaged in a build-up like the one that preceded *Operation Desert Storm* in 1991 and concentrated in rear disembarkation ports and airfields might be targeted. Similarly, power-projection forces and major logistics concentrations would be suitable targets for an attacker bent on disrupting a pending military assault.

During a BW campaign, large air bases and ports might be attacked to disrupt offensive and defensive air operations and re-supply efforts. The variety of BW delivery options would enable special operations forces to strike at strategic targets deep behind enemy lines. Allied populations in the crisis region could be threatened or attacked to undermine support for allied coalitions or coalition-building.

Daylight kills most established biological agents, so their impact might be less prolonged than that of more persistent chemical arms. However, replacing specialised personnel takes time, regardless of the condition of their equipment and infrastructure. Biological agents might thus be highly disruptive to military operations. In future, the biotechnology revolution might yield weapons that are more predictable and controllable, and thus more useful for tactical applications.

Biological-Weapon Proliferation

If proliferation trends are any guide, a number of countries believe that acquiring a BW arsenal, or the capability rapidly to develop one, serves their political or military interests. According to the August 1993 US Office of Technology Assessment report, 'Proliferation of Weapons of Mass Destruction – Assessing the Risks', countries suspected of having clandestine offensive biological-warfare programmes include China, Egypt, India, Iran, Israel, Libya, North Korea, Russia, South Korea, Syria, Taiwan and Vietnam. All have manned aircraft that could be used for BW delivery; and all but Vietnam manufacture or assemble ballistic and/or cruise missiles that could be configured for BW delivery. India claims to be working on submunitions for medium-range ballistic-missile applications. China could almost certainly develop submunitions, while Iran, Libya and North Korea, perhaps with various levels of outside assistance, could probably develop them as well.

Despite continuing inspections by the United Nations Special Commission on Iraq (UNSCOM) since the end of the Gulf War, the UN suspects that Iraq still retains biological arms. Iraq weaponised biological agents for delivery by *Scud*-type missiles prior to the War. None of this hardware has been uncovered, but UN inspectors found that Iraq had developed its short-range M-122 rocket for BW delivery. The weaponisation was rudimentary (about eight litres of liquid anthrax or botulinum in a unitary warhead), but each rocket could have contaminated a few square kilometres, perhaps more under favourable conditions. Iraq also filled 155m artillery rounds and R-400 free-fall bombs with biological agent and came close to completing development of aerial spray tanks.

Biological-Weapon Arms Control and Countermeasures

There are three principal tools with which to confront the BW threat: arms control; deterrence; and defence. No single component is adequate, and the overall approach is presently weakened by shortfalls in each area.

The principal arms-control instrument for stemming the spread of BW is the 1972 Biological and Toxin Weapons Convention (BTWC). The Convention prohibits the development, production and stockpiling of biological agents or toxins 'in quantities that have no justification for prophylactic, protective or other peaceful purposes'. As of July 1996, 138 nations had ratified the Convention, but its effectiveness has been undermined by the failure of member-states to establish a monitoring regime of any kind.

At its 1991 Review Conference, the states party to the BTWC created a group to draft a legally binding verification protocol, but detecting BW activities prior to full-scale development is a daunting challenge. The facilities required for BW research, development and production are small enough (perhaps requiring just 25m^2) to be contained within a legitimate pharmaceutical, medical, agricultural, food or fermentation facility. The dual-use nature of the facilities, equipment and materials used to produce BW agents makes distinguishing military from civilian activities extremely difficult. The number of personnel required to conduct a military programme is limited, which in turn limits opportunities to collect intelligence from human sources on clandestine activities. The Convention itself permits the development of biological and toxin agents in small quantities for peaceful and defence purposes. Such legitimate research projects and closely related commercial activities could, however, provide cover for an offensive biological-warfare programme.

A BTWC verification regime might have more success in seeking out such activities as testing, filling and stockpiling munitions, testing delivery systems, and conducting exercises to train military personnel in BW handling and operations. Detecting such activities would prompt closer scrutiny of the suspect nation's enterprise, at all levels.

With or without an effective verification regime, the BTWC has both established a norm to bar the possession of biological weapons, and provided an important symbol of most of the world's abhorrence of biological weapons and warfare. The wide consensus represented by the large number of BTWC members compels proliferators to acquire BW in secret, which adds costs, risks and time to their development programmes. Perhaps most important, the need for secrecy might slow the advancement of BW programmes from established agents to the acquisition of new, highly lethal ones produced by genetic manipulation. This could provide sufficient time to develop defences, and time is sorely needed. It was not until the 1991 Gulf War, and the discovery of Iraq's arsenal of weapons of mass destruction, that the allies began to increase their focus on nuclear, biological and chemical defence for regional contingencies.

A multifaceted defence against biological and other weapons of mass destruction is required to deter attacks and reduce vulnerabilities. The first line of defence remains deterrence by threat of retaliation. However, as a result of their adherence to the Convention, the industrialised countries, with the possible exception of Russia, have destroyed their BW arsenals and relinquished their ability to respond in kind. This leaves them with the threat of conventional attack or, in a few cases, the choice of nuclear retaliation. Whether such threats will indeed deter rogue states or terrorist attacks is questionable.

The allies' implicit nuclear threat may well have deterred an Iraqi BW strike during the Gulf War. However, the dependence of the US and other nuclear powers on nuclear weapons is declining. They have signed the Comprehensive Test Ban Treaty (CTBT), are withdrawing or eliminating categories of delivery vehicles, and de-emphasising nuclear training and readiness. As a result, nuclear response options will receive less emphasis than conventional retaliation in future regional conflicts.

It is increasingly understood that the threat of retaliation must be reinforced with a second type of deterrence – deterrence by denial. This requires acquiring a proven capability to defend against BW attacks, a capability that will deny an aggressor any political or military aims it might seek by using such weapons. Attacks that have little chance of success are less likely to be carried out. If deterrence fails, a denial strategy gives the capability to counter BW attacks effectively and provides an opportunity to defeat an aggressor through conventional means. Attack operations, as well as active and passive defences, are typically seen as integral parts of an effective countermeasures scheme.

Attack operations offer the highest pay-off of any approach, but they are also the most challenging to execute. The primary goal of attack operations is to seize, disable or destroy weapons of mass destruction and their launch platforms before the weapons can be used. Land, sea, air and special-operations forces could be used to destroy underground storage bunkers or production facilities, and to destroy aircraft and missiles prior to launch to ensure that their deadly cargoes are deposited on enemy soil.

Attack operations against launchers require timely intelligence and real-time, wide-area surveillance. To provide the latter, unmanned aerial vehicles that can loiter for more than a day while searching for suspicious facilities or machines are being investigated, as is using unattended ground sensors to detect and identify vehicles. Human-source intelligence is critical for identifying BW production and storage facilities; it would also be vital to authorities seeking to identify and pre-emptively attack terrorists bearing weapons of mass destruction.

Active defences would seek to destroy aircraft and missiles, or other BW delivery systems, as early as possible after their launch (i.e., close to or over enemy territory). However, the BW threat poses a critical challenge to defensive systems such as *Patriot* and other active systems – they might destroy inbound air vehicles and ballistic missiles, but disperse BW in the process, causing collateral damage. Passive defences seek to minimise damage from weapons of mass destruction that survive attacks and active defences. Passive defences include a wide array of measures and equipment, for example, respirators or masks to block BW particles, protective shelters including plastic-sealed rooms with filtered air, decontamination equipment, vaccination and antibiotics to prevent or treat infections by pathogens, and BW detection systems.

Preventive Measures

The NATO allies have in recent years increased their cooperation in developing countermeasures for biological and other weapons of mass destruction. But the development has not been well balanced, given the possible threat. For instance, with programmes that were well established even before the Cold War's end, ballistic-missile defences are advanced and could afford robust protection around the turn of the century. Defences against low-flying cruise missiles have been given less attention, and a critical shortfall undermines passive defence efforts: the inability to provide defenders with a rapid BW detection and identification capability. Without this capability, passive defence measures cannot be carried out in time. Protecting civilian populations presents a particular challenge.

Perpetual questions regarding the inter-operability of allied equipment also continue and are of special concern regarding weapons of mass destruction. Defences that are both unbalanced and unable to promote unity of effort ensure that serious vulnerabilities to biological-weapon attacks will remain. Unless such deficiencies are addressed vigorously, the next UN coalition to face an aggressor might collapse in the face of a BW threat to its forces or allied civilians.

Preventing determined proliferators acquiring biological and toxin agents appears to be virtually impossible. The complexities associated with weaponising and delivering biological and toxin agents might prevent large-scale attacks, at least in the near term. Nevertheless, these barriers are crumbling, and the revolutionary advances in biotechnology will probably remove them altogether in the first decade of the twenty-first century. While the BTWC has an important role to play in establishing a global legal norm against the possession of biological weapons, even with a verification protocol it is not a sufficient guarantee against acquisition and use. Effective defence requires a heavy investment in intelligence efforts and personnel protective measures, as well as maintaining the capability to carry out military attacks against development and production facilities, storage sites and delivery means. Unprotected civilian populations, however, will remain vulnerable to biological attack by an unscrupulous attacker. The industrial democracies must be prepared to defend their forces and populations by force if and when deterrence and diplomacy fail to prevent aggression by rogue states armed with biological weapons.

Secondary Boycotts and Allied Relations

Early in 1994, pulling together the strands of a policy that had been evolving for more than a year, US national security adviser Anthony Lake publicly set out a strategy for dealing with what he called 'backlash' or 'rogue' states. Writing in the influential US journal *Foreign Affairs*, Lake labelled Cuba, Libya, Iran, Iraq and North Korea as international 'outlaws' that merited not only the disdain but also the sanction and containment of the international community. Under Congressional pressure, the rogue-state policy steadily intensified during the first Clinton administration. In 1996, it exploded into a full-blown effort not only to contain or even topple such rogue regimes, but also to punish other states – even close allies – who deal with them. For an administration often criticised for indecision or lacking a grand strategy, the rogue-state policy, if not 'grand', was decisive.

It is not clear what effect this US policy has had on the states concerned; not even its proponents claimed that it would bring quick results. Elsewhere, however, it has had a rapid and clear effect. European (as well as Canadian and Mexican) governments have strongly opposed and vociferously denounced the US for this policy. America's allies vehemently disagreed with the logic of cutting off disagreeable regimes, arguing that openness and dialogue was a wiser strategy. Even more important, they resented foreign legislation that sought to punish their own companies that had business links with the so-called rogues. By the end of 1996, Europeans were retaliating in their own legal systems against the US laws, refusing to curtail their diplomatic or business dealings with the targeted states, and asserting their right to pursue different foreign policies from those of the United States.

The current transatlantic dispute about rogue states, and global policy more generally, is in some ways just another episode in a long series of transatlantic disagreements about NATO's 'out-of-area' policy that goes back to the 1956 Suez crisis, the Algerian and Vietnam wars in the 1950s and 1960s, and the US bombing of Libya in 1986. The current dispute is potentially as serious for two reasons. First, even more than in past crises, current transatlantic divisions undermine the effectiveness of either strategy – sanctions and containment are unlikely to bring down a regime when applied by the United States in isolation, but trade and dialogue are equally unlikely to foster change when the world's largest economy and political player is not involved. The result of a divided Western strategy towards rogue states, in other words, leads neither to the defeat of the unwanted regimes nor to their integration into civilised international behaviour, but rather to tensions within the Atlantic Alliance and an opportunity for rogue states to play the allies off against each other.

Second, and perhaps even more important, transatlantic divisions over rogue states and global foreign policy threaten the very cohesion of the Atlantic Alliance when that cohesion is no longer ensured by a common threat. Quarrels over 'out-of-area' issues never really threatened NATO during the Cold War because the need to stand together was always far more important than any extra-European disputes. Today, as NATO becomes more an alliance of common values than a defence pact, and when it occupies itself with issues that are no longer simply 'existential', external differences will matter more than in the past. If Americans come to see Europeans as free-riding appeasers of states that jeopardise global US interests, and Europeans come to see Americans as simplistic crusaders who seek to assert their authority unilaterally over their own allies, the unity of the Alliance and the willingness of Americans and Europeans to support each other in times of need will be seriously threatened.

Tightening the Screws

The details of Western policies towards each rogue state are different, but the general pattern is similar: the US urges containment and punishment, while Europe seeks change through dialogue and trade. The divergence in US and European policy is probably smallest concerning North Korea, both because Europeans are not leading actors in East Asia, and because it is the one case where US policy has been emphasising increased dialogue and trade rather than sanctions. While the US military containment of North Korea continues, Washington has been slowly moving towards engagement ever since Pyongyang agreed in October 1994 to freeze its suspected nuclear-weapons programme in exchange for food and energy aid and trade with the West. The US opening to North Korea is still tentative and uncertain, and it is perhaps not a healthy sign that it came about not because Pyongyang's behaviour was improving – far from it – but because North Korea's illicit nuclear programme had approached the point of becoming a truly dangerous menace. The message this experience sent to Iran and other rogue states – 'we will talk to you only when you have nuclear weapons' – was perhaps not what the United States had in mind, but it was given none the less.

Similarly, US and European policies towards Iraq have not yet led to serious, open disputes that could damage the Alliance, but only because the post-1991 Gulf War sanctions against Iraq allows the US to impose its policies of containment and isolation on sometimes unenthusiastic allies. It is not that anyone in Europe currently argues for dialogue and openness with Saddam Hussein or claims that Iraq does not pose a potential regional threat. But many in Europe accuse the US of continuing the confrontation with Saddam Hussein for domestic political purposes, and of 'moving the goalposts' concerning the lifting of UN sanctions. When, in September 1996, US air forces attacked Iraqi targets in response to an Iraqi incursion in the Kurdish fighting in Northern Iraq, many Europeans distanced themselves from the action. Only the UK offered full support, while France made its objections publicly known. Indeed France, a traditional ally of Iraq and the first Western country to break the post-Gulf War policy of ostracising the Iraqi leadership (when it welcomed Foreign Minister Tariq Aziz in January 1995), has been openly critical of US policy there, and Paris seems poised to promote and take advantage of an opening towards Iraq as soon as that becomes possible. A number of French and Italian construction and oil companies have already visited Iraq in anticipation of the eventual contracts that will be signed once the sanctions are lifted. Saddam Hussein's continued mistreatment of his own citizens and reluctance to cooperate fully with UN Security Council resolutions have prevented Europe from fully making the case for engagement, but there is little doubt that it is more prepared to do so than is the United States.

Far more openly contentious than Iraq is the case of Cuba, the first of the rogue-state problems to erupt into a full-scale legislative dispute between the US and the European Union. Long an area of policy disagreement between the EU – which is Cuba's largest single trading partner – and the US – which has maintained an embargo on trade with and travel to Cuba since Fidel Castro's 1959 revolution – Cuba policy became a high-profile issue with the 12 March 1996 passage of the US Cuban Liberty and Democratic Solidarity Act, known as Helms–Burton after its sponsors in the Senate. This legislation seeks to support the claims of US and naturalised Cuban-American citizens whose property was expropriated after the 1959 revolution. Title IV of the Act, which took effect immediately, gives the US government the right to deny entry into the United States to anyone who 'traffics' in such property. Even the family members of shareholders in a firm that uses Cuban sugar made in an expropriated plant, for example, could be barred from the US. (The first company to be affected by the law was a Canadian mining firm, Sheritt International, whose directors, including former Bank of England Deputy Governor Rupert Pennant-Rea, were banned from the US.)

An even more controversial aspect of the Helms–Burton Act is its Title III, which allows US companies to sue foreign firms and affiliates that have profited from the use of seized property. On 16 July 1996, President Bill Clinton announced that he was suspending implementation of Title III for six months in order to give foreign firms more time to comply with the spirit of the law. But the law still outraged the Canadians, Europeans and Japanese, who all condemned it as both unwise and illegal. The Canadian government immediately took measures to allow its companies to continue trading with Cuba, and the Europeans threatened both to disregard the US threats and to take the US to the World Trade Organisation (WTO). On 30 July, the European Commission approved draft legislation making it illegal for EU companies to comply with the Act and giving them the right to counter-sue in European courts. European officials denied that Cuba posed any threat whatsoever to international security, and argued that rather than tightening the US embargo by threatening its allies, the embargo should be dropped altogether.

Among all of the rogue-state disputes, the most consequential concerns Iran. Since the 1979 Islamic Revolution and the seizure of US diplomatic personnel held hostage in Tehran for 444 days, Iran has been a particularly determined adversary of the United States – 'the Great Satan' to the adherents of the Islamic regime. (The US negotiator during the hostage crisis was Warren Christopher, who showed no signs of having forgiven his former antagonists when he became Secretary of State 15 years later.) The US cut off diplomatic relations with Tehran in 1980, and Iran is consistently placed on the State Department's annual list of state supporters of terrorism. US leaders have accused Iran of threatening its neighbours, seeking to sabotage the Middle East peace process by backing anti-Israeli groups like *Hizbollah* and *Hamas*, pursuing an illicit nuclear-weapons programme, and murdering Iranian dissidents abroad. Europe – which does not deny the Iranian regime's wrongdoings, but thinks that the US blows them far out of proportion – has sought to influence Iran through a process of 'critical dialogue' and continued trade. The US, however, has been resolutely determined to isolate and punish Tehran. Washington has scarcely hidden the fact that, through its policy of 'dual containment' of Iraq and Iran, it would be happy to see the demise of both regimes.

Despite this long antagonism, it was not until the beginning of 1995 that the Clinton administration began a more active campaign to contain or even destabilise the Islamic regime. After Iran signed a $1bn contract with Russia in January 1995 to build two nuclear reactors in Iran, and Iranian leaders appeared to sanction Palestinian terrorist attacks on Israeli civilians the following March, Washington decided to end the inconsistency that condemned other countries for trading with Iran but allowed US companies to do so. As late as 1994, US companies were the largest purchasers of Iranian oil, and US exports to Iran amounted to more than $1bn. Importing Iranian goods into the United States was banned in 1987, but 'indirect' trade, whereby US companies bought and sold Iranian goods elsewhere, was allowed. On 30 April 1995, President Clinton announced a ban on all US trade with and investment in Iran, and forced the US company Conoco to drop the oil-exploration deal it had signed with Iran the previous month. Going even further, in July 1996 the US House of Representatives followed the Senate in passing the D'Amato Act, a bill, like Helms–Burton, that would punish even foreign companies that did business with Iran. The Act, sponsored by New York Republican Senator Alfonse D'Amato, obliges the President to implement at least two of six possible sanctions, ranging from an import ban to lesser penalties, on all companies that invest more than $40 million in either the Iranian or the Libyan oil and gas sectors. President Clinton signed the D'Amato Act into law on 5 August 1996.

Because of the much higher levels of trade and investment at stake, Europe's reaction to the D'Amato Act was even more vehement than it had been to Helms–Burton. European leaders argued that it was wrong to try to change Iran by crippling its oil sector, and that such measures would actually prove counter-productive by giving the Islamic regime an excuse for its poor economic performance, and by making the West seem hostile to Islam. More specifically, US allies protested against the principle of 'extra-territorial' legislation – the idea that Washington can impose its own laws on foreign companies and countries – that was common to Helms–Burton and D'Amato. EU trade commissioner Sir Leon Brittan called both US Acts 'objectionable in principle, contrary to international trade law and damaging to the interests of the European Union', and the EU began to apply similar measures against D'Amato to those it took against Helms–

Burton. If the Iran and Cuba cases are brought before a WTO dispute panel, the outcome would be of enormous significance. The US would almost certainly appeal to WTO provisions that allow countries to suspend international trade rules to protect 'essential security interests', while Europe would counter that national security was not at stake. If the WTO found in favour of the US, the precedent of appealing to national security could undermine the whole WTO system; if it found against the US, American support for the Organisation could be undermined.

The transatlantic debate about Libya mirrors in many ways that about Iran. Like Iran, Libya is seen in the United States as one of the world's foremost supporters of state-sponsored terrorism, and Libya's leader, Moamar Gadaffi, is as reviled in the US as are any of Iran's leaders. Europe does not exactly support Gadaffi's regime, but it opposes trade sanctions against it, just as most European states (with the exception of the UK) opposed the punitive air-strikes on Tripoli made by the US in April 1986 in response to what Washington claimed was evidence of Libyan-supported terrorism. Libya, unlike Iran, is the object of sanctions imposed by the UN after Gadaffi's unwillingness to hand over two Libyans accused of bombing Pan Am flight 103 in December 1988, but the UN sanctions only cover air travel to and from Libya, and not trade or oil development, which Europe opposes. Frustrated by Europe's unwillingness to support harsher sanctions on Libya, the US Congress in June 1996 enlarged the D'Amato Act, originally designed to apply only to Iran, to include Libya.

The Sources of Transatlantic Divisions

Why do the US and its allies have such divergent views about rogue states, and why are their policies so different? At the heart of the dispute is a genuine difference in policy analysis between the US and Europe about how best to deal with, and eventually change, objectionable regimes. Europeans argue that dialogue and trade are the best tools of leverage, and that prosperity and human contact are ultimately more successful means of spreading democracy and cooperation than anything else. Trade, in this view, fosters common interests and greater harmony among nations, and stimulates business activity and political change. Europeans appear convinced, in any case, that US policies of punishment and containment simply do not work, and that demonising élites in rogue states will only increase their determination to resist Western hegemony and give them a public excuse for poor economic and social performance.

The US responds that the European view is either cynical or naive. The leaders of rogue states are probably unreformable, and even if it were possible to change their attitudes over time, they would do so much damage in the meantime that the risk of trying is too great. Trade and dialogue with rogue states does not foster openness and political change, but simply gives the hostile regimes more resources with which to carry out their destructive programmes, taking away any incentive they might have to amend their ways. The US objective is thus to maintain the pressure and force these states to change, while preventing them from doing much damage in the meantime.

It would be misleading to conclude, however, that the transatlantic dispute over rogue states is simply a difference in analysis, and that the respective positions on the issue could just as easily be reversed. In fact, the US and European positions have as much to do with their own domestic politics, economic interests and political cultures as with a genuine policy dispute.

It is hard to deny that domestic politics have played a critical role in both US and European policies towards Cuba, Iran, Iraq and Libya over the past several years and even decades. Cuban-Americans have long been influential in US domestic politics, and it is certainly no coincidence that the Helms–Burton Act was signed into law by a President known to be uncomfortable with it just four months before a presidential election. The states of Florida and New Jersey, where many Cuban-Americans live, have between them 40 electoral votes (nearly 15% of the 270 needed for a presidential victory), and it is unrealistic to argue that Cuba, especially after the end of the Cold War and the death of global communism, poses a strategic threat to US interests. The 'strategists' of Helms–Burton would have a hard time explaining why Fidel Castro and his regime are a greater threat to human rights or US interests than many regimes around the world with which the US maintains close relations.

Iran does pose potential threats to US interests, and certainly mistreats its own citizens, but there also seems little doubt that those threats are often exaggerated in the US. The D'Amato Act faced an uncertain fate in Congress until the bombing of a US military base at Dhahran, Saudi Arabia, on 25 June 1996 and the explosion of TWA flight 800 on 17 July made it seem imperative to 'get tough' on Iran, even though there was no solid evidence that Tehran had anything at all to do with either incident. Taking tough action against Iraq and Libya – whose recalcitrant leaders are seen by many Americans as responsible for past acts of international terrorism – is also popular in the US, which has little to lose by doing so as it has few economic interests in either country.

Europeans, of course, do a lot of business with all of these 'rogue' states, and that is the main reason why their domestic politics lead them in a different direction. Whereas the US has not traded with Cuba for more than 35 years, for example, the EU and Canada are the source of more than 70% of Cuba's imports, valued at over $1.3bn per year. European trade with Iran and Libya is even greater. The EU's annual trade with Iran and Libya is more than $20bn and the two countries account for more than 20% of EU oil imports. Some large European companies (such as France's Total and Elf and Italy's Agip) have invested hundreds of millions of dollars in both countries. Iran also has debts with Germany of more than $8bn. Europeans would surely be more willing to sanction Iran if they did not have such significant economic interests at stake.

Another reason why the US takes a harder line on rogue states than does Europe is that Americans see themselves as, and in fact are, more responsible for containing threats to international order than Europeans. If a rogue state challenges regional or global security, the United States takes the lead in dealing with the problem, and it has often come into conflict with those states that have challenged the status quo. It is thus not a coincidence, or even solely due to geography, that Europe trades more with Cuba, Iran, Iraq and Libya than does the US. These states have all directly challenged Washington's global leadership – sometimes through propaganda, sometimes through terrorism and sometimes through war – and thereby provoked far more US hostility towards them than European hostility. It was the US that was targeted by Soviet nuclear weapons in Cuba during the 1962 missile crisis, the US whose diplomats and other citizens were seized by Iranian 'students' for over a year during the Islamic Revolution, and the US that led the international coalition against Saddam Hussein; all of these episodes seared American political consciousness and their legacy endures to the present day. The situation today is not altogether different from during the Cold War, when the US was the self-appointed (and widely accepted) leader in the fight against communism, resulting in what the Europeans often considered excesses in Central America, Asia and even in Europe itself. Today, just as then, some Americans believe they are responsible for coping with challenges to regional or international order, and that Europeans are able, and all too willing, to 'free ride' behind them.

A third and final reason for the greater US hostility to these states is that US political culture seems far more susceptible to 'demonisation' and differentiation between 'friends' and 'enemies' than European political culture. Perhaps because of their much longer and more complicated histories, in which all states have at various times been 'good', 'bad' and both at the same time, Europeans seem more willing (in the eyes of some Americans all too willing) to take a moderate view of other states' behaviour. Americans, on the other hand, with their idealistic (some Europeans would say naive) vision of their own place and role in the world, have always been readier to divide states and causes between good and evil. US wars have thus been fought for causes as grand as 'making the world safe for democracy', 'defeating fascism' and 'creating a new world order', rather than for such crass goals as national interests. All of this leads the US to view rogue states as not merely a little more problematic than others, but as sinners who must be reformed. Europeans are more prepared to live with what they believe they cannot change.

Looking Ahead

The deep-seated European–US differences in domestic politics, economic interests and political culture make the rogue-state dispute far more difficult to resolve than it would be were it simply a matter of policy analysis. As a result, and since there is little sign that the problem will disappear soon, the chances that these differences will continue to be an irritant to transatlantic relations – or perhaps even worse – are very great indeed.

Taken in their purist form, it is difficult to see which argument has more merit. It must be said that neither strategy for influencing hostile regimes has an impressive track record, and neither 'relentless containment' nor 'critical dialogue' is without its flaws. The US would have to admit, for example, that while its policy has made life more difficult for the governments and people of Cuba and Iran, neither 38 years of sanctions against Castro nor 17 years of confrontation with Iran has undermined either regime's hold on power (the opposite might in fact be true). Indeed, the historical record for changing regimes through sanctions is exceedingly poor. Whereas the US might argue that sanctions would have worked better had they been applied universally rather than just by the US, the record from Iraq – where universal sanctions do hold – is also not good: the country is destroyed and the population suffering, but the regime survives. The international track record of changing authoritarian regimes through engagement and economic development seems much better, as the US itself has implicitly admitted in its own, selective pursuit of contact with many unappealing leaders. There is thus apparent inconsistency in the US approach – it seeks to influence Chinese and North Korean behaviour through trade, dialogue and engagement, but denounces those methods when applied to Cuba, Iran and Libya.

Where the US clearly has a point, however, is in arguing that change through engagement takes a long time, and that rogue regimes have the potential to damage regional and even international security, as well as their own people, in the meantime. Perhaps, if left to trade and interact in the community of nations, Iran or Libya would one day develop into status quo or even liberal powers; but what should the world do in the meantime if they are murdering their own citizens at home and abroad, developing weapons of mass destruction, seeking to sabotage Arab–Israeli peace and issuing death edicts against Western authors? Money is fungible, and every dollar earned by rogue regimes can be put to use by them to maintain their grip on power and to pursue their interests abroad. Europe's critical dialogue might appease European consciences and sound good in theory, but there is little evidence of it influencing rogue states in what the West would consider a positive direction.

Because the issue of how best to deal with rogue states is inconclusive in the abstract, policies will doubtless continue to be based on national interests and perceptions, and will therefore continue to differ across the Atlantic. There might, however, be something to be gained in efforts to develop a common strategy, whereby senior US and European officials would seek to identify areas of 'unacceptable' policy and agree what should trigger sanctions. These areas might be concrete proof of terrorism (evidence of which the US would have to share with the Europeans, rather than simply assert) and the pursuit of weapons of mass destruction. If taken seriously on both sides, such an effort could also protect political leaders from domestic criticism if the US and EU together decided on policies one of their publics, business communities or legislatures did not like. The Clinton administration could refuse Helms–Burton-like legislation, for example (which it was known privately to oppose anyway), on the grounds of Alliance cohesion, arguing that unilateral policies would not work. This was not altogether different from the administration's stance in 1995 on lifting the arms embargo against the Bosnian Muslims – the administration opposed Congress, arguing that Europe would have to agree for it to work. European leaders could also cite the need for transatlantic consensus in applying sanctions on a rogue state that their business communities might oppose. Such efforts, or even the creation of a formal Euro-Atlantic Council for the purpose, would not ensure transatlantic harmony or convergence of views (or interests), but it could at least foster a better understanding of each side's arguments and provide a measure of political cover for leaders whose views are not as extremist as those of some of their legislators.

Ultimately, of course, there is no simple institutional solution for the problem of rogue states and transatlantic relations. But if the Atlantic allies cannot find a way to agree on policy, they must at least understand the negative consequences of failing to do so. By not appreciating the importance to the US of constraining states that threaten its interests around the globe, Europe may be undermining US support for it at a time when such support is no longer guaranteed. And by seeking unilaterally to impose its will on dissenting allies, the US might well be provoking the very sort of independent European foreign policies it is seeking to avoid. None of this is very healthy for a transatlantic security alliance whose only truly common threat has disappeared.

Europe's Post-Communist Democracies

The large geographic area previously occupied by the Soviet imperium has demonstrated neither a collective great leap forward, nor the general cataclysm predicted in some quarters, but rather a steady diversification of the political, economic and security prospects of its 20 constituent states and emerging regions. Among the countries of the former Soviet Union and the Warsaw Pact, there are definite winners (Poland, the Baltic states and the Czech Republic, for example), which share historical, geographical and even cultural advantages and common policy preferences. There are, however, also definite losers (Bulgaria, Belarus and Slovakia, for example), states which are falling behind their neighbours, but each in its own way. The vast majority of the former Soviet and Warsaw Pact states, including Russia, Kazakstan, Ukraine and Uzbekistan, lie somewhere in the middle. They are moving towards political and economic transformation, but have not yet secured the results already enjoyed by the leader states. It is the fate of those in the middle, especially Russia, that will determine whether the transition from communist to post-communist and beyond has been successful.

The increasing diversity of Europe's post-communist space and the new political and economic patterns that have already emerged are best understood by examining the five forces currently transforming this once-unified geographic area. These forces include the power of sovereignty, political transformation, economic stabilisation, unequal security prospects and diverse international ties.

The Power of Sovereignty

Although the desire to establish and maintain full sovereignty was never in doubt in the non-Soviet Warsaw Pact countries, many observers did question the capacity of the former Soviet states to create and maintain a sovereign existence. Yet, since the disintegration of the Soviet empire in 1991, sovereignty has proved to be a durable commodity for all but a handful of regimes, which are torn by civil strife or challenged by secessionist movements on behalf of ethnic or regional enclaves. Even these movements do not seriously question the existence of a sovereign Azerbaijan, Georgia, Moldova or Russia, but instead wish to create new sovereign entities to co-exist with these states.

The existence of these 20 new states is an international fact, no mean feat considering that less than half had any prolonged experience of independence to draw upon. Moreover, in defiance of the worst fears, the success of the majority of these states is not due to unleashed nationalism. While much feared as an irrational and suicidal impulse for the myriad ethnic groups in the former Soviet Union, nationalism has been, by and large, a much weaker force than was predicted. It has clearly played an important role in ethnically homogeneous Armenia or in the political movements of the Baltic states, particularly Estonia and Latvia. Slovak nationalism has also been a potent force in creating and sustaining a new state, although it is also a prime cause of Slovakia's fall from the inner circle of Central European states deemed ready for membership in NATO and the European Union. Former Georgian President Zviad Gamsakhurdia's extreme nationalism did not suit multi-ethnic Georgia and spurred the restiveness of Ossetia and Abkhazia.

In Ukraine, Kazakstan and Uzbekistan, nationalism has been exaggerated as a political force. Genuine nationalist feelings are developing in the new states, but these are of long-term, rather than immediate, significance. Most regimes in states with large ethnic minorities have engaged in a subtle balancing act between satisfying the demands of the titular nationality and reassuring large ethnic minorities, especially the key Russian populations of Ukraine and Kazakstan. In Central Europe, the September 1996 Hungarian–Romanian Treaty represents a potential breakthrough in the thorny issue of the Hungarian minority in Romania. However, the continued problems in Hungarian–Slovakian relations, despite a similar treaty in 1995, should prevent premature optimism that the problem of ethnic and cultural minorities in Central Europe has been fully resolved.

For those countries with little previous experience of statehood, multiple and even contradictory sources of support for sovereignty have emerged. The most important was the collapse of the imperial power itself. The Soviet Union was unable to maintain its hold over the satellite nations of Eastern and Central Europe in 1989, or to quell rising national and regional challenges within it. The power of its main successor state, Russia, has been radically constrained and transformed by its own internal problems and reforms. It can neither coerce nor materially support integrative forces in the rest of the area. Instead, the new states must function on their own, precisely because no 'big brother' is available, even if one were desired.

By choice or by default, the new capitals have become the distribution centres of money and power, even if on a much reduced scale from that of the old Soviet Union. Key regions and industries which formerly turned to Moscow now take their case to Almaty, Tashkent or Kiev. Sovereignty has also brought material and political benefits to the ruling élites of these countries. Many have found great wealth and political influence far exceeding what they would have received as regional leaders in the old Soviet Union. Sovereignty also brings for the first time control of the state's natural resources, although the absence of pipelines and other infrastructure prevents these resources being turned into export revenues quickly. Sovereignty places national languages and cultures above or on an equal footing with Russian, and restores direct links with those European and Islamic lands once mediated by Moscow.

While Russia remains a key state in the region – and one that is likely to grow in importance as it recovers its political and economic health – the likelihood of a wholesale reversal of the trends that favour sovereignty over an integrated Russian-led community has been consistently overestimated by analysts. There are integrative tendencies within certain parts of the former Soviet Union, but these are united border regions or countries that share an important transportation corridor. These trends are emerging as the result of ordinary economic activity and self-interest not imposed from outside. Some states seek close ties with Russia while others are quite dependent on it. In the former category, Armenia, Georgia, Kazakstan and Kyrgyzstan see Russia as a support against more serious threats within the immediate neighbourhood, but at the same time they work to minimise blatant Russian influence over their politics and economic activities. Tajikistan and Belarus, on the other hand, are clear examples of failing regimes in need of Moscow's support. But an integrated community of weak states is like the stone soup of the children's tale: it provides little nourishment unless others can be induced to provide the meat and vegetables that make real broth.

Although sovereignty is strong, the states themselves are still weak. This weakness is apparent in tax collection, the regulation of commerce, control of national borders and the fight against crime and corruption. It is also evident in the wide range of regional and even secessionist challenges to existing regimes. Secessionist movements in the Transdniestr, Abkhazia, Nagorno-Karabakh and Chechnya have already achieved *de facto* autonomy from their governments. The only similar movement in Central Europe was a non-violent one, leading to the establishment of separate Czech and Slovak Republics in January 1993. While there has been no second round of the 'parade of sovereignties' that some feared might emerge in Russia, Ukraine and elsewhere in the former Soviet Union, regionalism in all these states has become a potent force, drawing political power and economic influence from the centre. Within Russia, Tatarstan, Tuva and other subjects of the Federation have attained a high degree of autonomy simply by bargaining with a weak central government.

Political Stability

The 20 post-communist countries can be grouped into four broad categories:

- regimes with functioning democracies and robust civil societies;
- regimes defined more by pluralism than by established democratic institutions and strong civil societies;
- regimes that place order above Western-style democracy; and
- unstable regimes.

An increasing pluralism of political, economic and cultural life is apparent throughout, yet this pluralism in many cases still seeks a stable, enduring political and societal framework. There is no universal agreement in the post-communist states on the political and societal model that would provide this framework.

There is little doubt that Poland, the Czech Republic and Hungary are stable democracies with the Baltic states of Estonia, Latvia and Lithuania not far behind. These countries have strong state institutions, multi-party systems, free-and-fair elections and, despite decades of communist rule, robust civil societies. In these countries a social consensus exists which supports a thoroughgoing transition from communism, democratic reforms and free markets. Of course, they have their problems. The Polish military is still not under effective civilian control, and Estonia and Latvia have created legal obstacles to the easy integration of their ethnic Russian minorities into political life. Although by no means the oppressive human-rights issue trumpeted by Moscow, the way they have dealt with Russian minorities reflects an uneasy compromise between full democracy and the protection of the state's national and ethnic character. Former communist and socialist parties have returned to power in Hungary, Lithuania and Poland, but their return has not altered basic domestic or foreign policies. The Polish left, for example, in sharp contrast to its past, embraces NATO membership and free markets. What is common to all of these states is a presumption of belonging to a European civilisation and a corresponding political horizon shaped by European political and economic norms.

The second category of states is a diverse group. Some have made great strides in political reforms, others continue to rig elections. In all of them, old and new forms of political life exist side by side, and it is not uncommon to find new business and political interests founded on the basis of old Community Party ties. In Russia, civil society was in tatters, but the state tradition was strong. Ukraine, Kyrgyzstan, Moldova and Armenia started from a weak basis for both a democratic state and robust civil society, yet have made great strides. It is more difficult to characterise Georgia, which has done much to re-establish political order since the election of Eduard Shevardnadze in 1992, yet where the central government still has little or no power at all in key sections of the country. Slovakia has taken a nationalist detour that may eventually lead back to Europe, and Bulgaria has also taken a detour, but one inspired by the persistence of the old ruling party and disastrous economic policies. Of all the Central European states, Romania was the most disadvantaged by the ravages of the Ceausescu regime. Its transition from the old regime was the most violent, and its first non-communist government represented far more continuity with the rejected past than the other Central European regimes that emerged in 1989. Yet the November 1996 election of President Emil Constantinescu, a reformist figure, raises real hope for the future.

These states have become genuinely pluralist. Tight police controls have disappeared, although there continue to be irregularities as strong courts and legal protections are still not the norm. The press is free, if often weak and uninformed. In Russia, the press is more developed as an institution, but its performance during the 1996 presidential elections as an uncritical supporter of President Boris Yeltsin's re-election shows how far Russian journalism still has to go. New economic and political interests have appeared, with the old nomenclature and 'new money' having greater influence than other interests. Elections are contested, even if not all have been free and fair. In Russia, elections have become an accepted and hotly contested part of political life. The same is true in Ukraine, where the 1994 presidential elections unseated an incumbent president and power was transferred without incident. The Moldovan electorate also turned out an incumbent president in December 1996, and while Armenian elections in September 1996 returned the incumbent president, they were judged a sham by foreign observers. However, in these countries, a weak civil society, poorly defined political parties, and the new-found advantages that have accrued to the major political winners (usually insiders) continue to distort the day-to-day process of governing. Even in the most advanced of these states, strong oligarchic structures dominate key political and economic decisions.

The third group of regimes rejects democracy as a near-term goal, defining themselves instead as enemies of disorder. The regimes of Turkmenistan, Uzbekistan, Azerbaijan and, increasingly, Kazakstan are defined by the concentration of executive power. Their leaders speak of their fidelity to Asian values and native traditions, and all have engineered not only their rise to power, but also the extension of their terms in office, some for life. Weak legislatures are the norm, and elections, where they occur at all, are riddled with fraud. Turkmenistan's President Saparmurad Niyazov has gone the farthest in preserving old-fashioned socialist controls and even the cult of personality, but all of these regimes are authoritarian in spirit and, where state institutions are strong enough, in practice as well. Yet there is real doubt about whether these regimes are as a strong or as stable as they appear. The basic state mechanisms may be repressive in intention, but are not quite up to the task they have taken on. None of these regimes has faced a leadership transition, which will put great strain on the basic state institutions as the strong leader passes from the scene leaving his heirs to battle for supremacy.

The fourth group is made up of unstable states. At present, these include Tajikistan and Belarus, although others on the verge of economic failure, unable to resolve existing and potential regional conflicts or sustain a peaceful transition of power, could also join. Tajikistan's President Imomali Rakhmonov aspires to authoritarian control, but his feeble state is in the midst of civil conflict. Belarus' Aleksandr Lukashenka has succeeded in consolidating his power, effecting a wholesale change in the state's constitutional structure in November 1996. Yet the power he has gathered has meant little for the failing Belarusian economy. Lukashenka's long-term hopes are pinned on integration with Russian (and subsidies from Moscow). His lasting accomplishment, however, is likely to be to make Belarus into the poorest and least stable state in Central and Eastern Europe.

Economic Prosperity

The collapse of the Soviet system and then of the Soviet Union itself forced all the countries of the region into a rude encounter with economic reality. All have had to deal with spiralling prices, slumping production and a declining gross domestic product (GDP), and have had to choose between real economic reform or continuing, under much less favourable conditions, old-style governmental controls. The great diversity of these states' economic prospects today is by and large the result of the path each has chosen.

It is clear that, except for a handful of states in economic crisis, most of them have made at least modest strides towards a market economy. The states quickest to carry out sweeping reforms include Poland – now the fastest-growing economy in Europe – the Czech Republic, Hungary, Armenia and Estonia. Latvia has lagged behind because of a banking crisis in 1995, although it too is undertaking serious reforms. Romania has demonstrated very high rates of growth in the past two years, but this has not been accompanied by substantial restructuring and thus may not be a good indicator of future economic health. Russia has created a stable currency, liberalised its economy and substantially completed the work of privatisation, leading many experts to believe that it is on the verge of sustained growth in 1997 or 1998. Ukraine lags farther behind, particularly in the privatisation of state assets and in state dominance of GDP, but President Leonid Kuchma's government has made impressive gains in controlling inflation, price liberalisation and introducing a new currency, the *hryvna*. Moldova, too, has made progress, and its private sector now accounts for 60% of its GDP.

The losers include Belarus, Tajikistan and Turkmenistan, some of the most illiberal economies on earth. Bulgaria has also fallen into the losing camp, with its disastrous slide beginning when it broke away from serious reform in the early 1990s. The test in the immediate future is whether a new president and the old socialist government can rediscover the courage to conduct reform or risk sliding further towards economic chaos and civil unrest. Taken together, the economies of Europe's post-communist states make a strong case for the efficacy of radical economic reforms – financial stabilisation, liberalisation and privatisation – and the failure of go-slow approaches or 'third ways'.

Like Bulgaria, Albania is going through a period of serious civil unrest. Also like Bulgaria, the root of its current political troubles is economic. Following the fall of communism, Albanians looked forward to a capitalist prosperity they believed was just around the corner. Corrupt businessmen, however, took advantage of these expectations, creating investment funds that promised large monthly returns of over 20% in the days before their collapse. These funds attracted the savings of perhaps a third of Albania's population. They were little more than pyramid schemes, which finally collapsed in December 1996. The anger of the Albanian population spilled out into street protests, which turned violent in Vlora and other southern Albanian towns in March 1997. The violence in the south led to the death of 14 Albanians, the resignation of President Sali Berisha's government and the declaration of a state of emergency. The President himself appears to have ties to at least one of these pyramid schemes. The Albanian Army is already involved, with the prospect of greater violence and instability tearing at the still-fragile structure of Albanian society.

Russia, Ukraine and other economies in transition, however, reveal certain persistent pathologies in the reform process that limit the scope, speed or even perceived legitimacy of reform. These include widespread corruption, the continued existence of a Byzantine bureaucracy that only works when bribed, weak civil and commercial codes and the absence of effective court systems to enforce them, as well as arbitrary tax policies that appear to exclude rich and powerful indigenous interests. During the transition to market economies a handful of government officials, bankers and entrepreneurs have parlayed their insider connections to fabulous wealth, while wages and pensions in key sectors go unpaid for months at a time. These distortions will continue to shape the politics and emerging markets of these struggling states.

Security 'Haves' and 'Have-Nots'

When it comes to security in the post-communist states, the new regimes have not been treated equally. All have benefited from the disappearance of the military confrontation between the Soviet Union and the West. The former Soviet nuclear forces are no longer aimed at Western or Chinese cities and no conventional force capable of large-scale offensive action exists within Europe or the former Soviet Union as a whole, nor is one likely to emerge in the next decade or beyond. The Soviet military machine has been broken into the new militaries of 12 of the successor states, but the pieces of this former machine are by no means equally distributed. The largest have gone to Russia and Ukraine, even though the armies of both these states are in complete disarray, with neither now inclined towards large-scale weapons acquisitions or the build-up of forces. Despite this positive trend, the western and southern portions of the region face very different security futures.

In the west, the basis for a regulated security environment is already in place. National militaries remain small and none of the countries is planning to enlarge them soon. A series of arms-control agreements regulates nuclear and conventional force levels and the behaviour and transparency of forces in the field. As national militaries are created or reformed in Europe and the western regions of the former Soviet Union, a major opportunity exists to preserve and stabilise the current low levels of military force, especially in the context of renewed Conventional Armed Forces in Europe (CFE) Treaty negotiations. The greatest immediate obstacle to maintaining this stable military balance is the potential for renewed friction between Russia and the West over NATO expansion. Indeed, the greatest challenge in European security policy in the next couple of years is likely to be to prevent misunderstandings over NATO from escalating into a series of military measures and countermeasures. Barring such a development, there appears little reason to expect the re-emergence of any major military confrontation in the region.

There are, however, sources of regional tension. The conflict between the Moldovan government and the Transdniestr, complicated by the presence of the Russian 14th Army, could re-ignite. Instability in Belarus or in Crimea could also generate a crisis. However, in the western portion of the post-communist space, the greatest danger comes from the leakage of military – particularly nuclear – hardware, know-how and technicians to rogue states or regions of conflict. Security analysts fear for the safety and security of Russia's residual nuclear force far more than they fear Russian nuclear doctrine or intentions.

The situation is more complicated in the south. No overwhelming military force threatens any of these states, but the war in Afghanistan does present an imminent source of instability to several of the states. Moreover, the weakness of existing governments has created a dangerous military environment, one of low intensity and regional conflicts combined with a fluid situation of illicit trading in arms and ammunition. The weakness of the Russian and indigenous military forces and the difficulties of terrain give great scope to paramilitary, guerrilla and other irregular forces. The weak states themselves, often governed by unstable coalitions of political groups, each with its own armed forces, create incentives for the use of force to resolve political conflicts. Existing or smouldering conflicts in Chechnya, Nagorno-Karabakh, Abkhazia, Tajikistan and near-by Afghanistan give a strong sense of what the future could hold in these and other southern locations. Such distinct security prospects for the west and the south will unfortunately diversify the prospects for stability and prosperity in the nations that have recently emerged.

Returning to the Outside World

The communist bloc was closed from normal interaction with the outside world. By contrast, the countries that have since emerged are open to such interaction. The return of open borders and genuine frontiers is every bit as radical a step as democratisation or major economic reform. It is forging new trade and communications links, no longer routed through Moscow.

New pipelines are planned, and Russia is understandably nervous that these pipelines will bypass its network, robbing it of revenue and influence. The new oil- and gas-producing states – chiefly Kazakstan, Turkmenistan and Azerbaijan – foreign oil and gas companies and outside customers want alternative routes to ensure their products reach their markets and are not subject to the economic and political control of a single country. A range of projects are under discussion, from new links to Russia's existing pipelines to wholly new routes that bypass Russia altogether. Kazakstan has agreed to ship its Tengiz oil through a new connecting line to the existing Russian system. A Caspian Sea consortium, which includes Russian oil companies, has agreed on a dual pipeline system for moving the early oil from Azerbaijan's Caspian wells to market. One route runs north from Baku through Grozny to Novorossiysk and the existing Russian pipeline system. A second route would ship oil from Baku to the Georgian port of Batumi. This dual system is the first stage in a fierce struggle for alternative routes. Turkey wants a new pipeline for this later oil constructed to one of its Mediterranean ports. Iran offers an attractive alternative route, although the current US embargo will postpone its development for the foreseeable future. While it is difficult to predict which of the many ideas for new pipelines, some involving China, will eventually be realised, the trend towards supplementing the existing Russian pipeline system with alternative routes is clear and is likely to strengthen (see map, p. 255).

All around the periphery of the former Soviet Union, new business, cultural and diplomatic links have been formed, and the Chinese have become a major presence in Central Asian trade. Russia is now, according to some estimates, Turkey's largest trading partner, and in turn Turkish businesses are a ubiquitous presence in the construction industry and consumer-goods trade throughout the former Soviet Union. This trend will only continue, as the natural resources and selected technologies of Russia and the former Soviet Union grow in importance for China and the dynamic and energy-poor economies of Asia.

Of greatest long-term concern is the potential for diplomatic friction or even long-term conflict created by the return of the outside world to this once-isolated area. Russia has underscored its deep concerns about enlarging NATO to include Poland, the Czech Republic and Hungary. This expansion will not only push NATO's boundaries to the eastern Polish border, but will also expand its field of interest in essence to the whole of continental Europe. Moscow has voiced similar concerns at various times about Turkish or 'Islamic' influences in the south. In the east, Russia must cope with the expansion of Chinese power. Each of these areas could become a source of inter-state friction and strategic surprises as the assumptions of the old game give way to new players and a new set of rules.

The Implications of Differentiation

The states of the former Eastern bloc have now become highly differentiated in their politics and prospects. In the context of the debates on NATO and EU enlargement, Commonwealth of Independent States (CIS) integration, and continued Balkan fragmentation, new foreign-policy attitudes will both manifest themselves and probably lead to more inter-regional tensions. Certainly the relations between the 'ins and the outs' of the European integrative process are bound to be strongest amongst the new entrants to the game of Western institution-building from the East. NATO chose to enlarge in part to avoid the nationalisation of security policy in Eastern Europe, to avoid the competitive alliance-building among Eastern European states and to maintain a multilateral approach to security. But as the gap between the successful states and those excommunicated from the Western political church widens, the possibility for regressive politics by the losers will increase. Managing the security dilemma caused by the disenfranchised will be a greater security task than maintaining the stability of those who, through that stability, will have been allowed earlier entry into the Western club.

Arms Control: A Mixture of Good and Bad

The long effort to stop nuclear testing culminated in September 1996 when an overwhelming majority of the world's nations signed the CTBT. It will be a long time, if ever, before the Treaty enters into force, however, partly because of India's adamant opposition. Nevertheless, the support the Treaty has received from almost all other countries means that a ban on testing is likely to take effect even without full ratification. Combined with major reductions in deployed nuclear weapons, and a renewed and invigorated nuclear disarmament debate, the vote in favour of the CTBT has given the disarmament process new impetus.

If 1996 was a good year for some aspects of nuclear disarmament, it was a poor one for the Chemical Weapons Convention (CWC), the BTWC, the second Strategic Arms Reduction Treaty (START II) and the Anti-Ballistic Missile (ABM) Treaty. In each case, domestic political considerations prevented any real advancement on these vital areas. Agreement was reached in May 1996, however, on the CFE Treaty's flank issue, although these changes had still not gained respective parliamentary approval by the end of 1996.

Bilateral Efforts

The Strategic Arms Reduction Treaties

START I and II took very different courses in 1996. In its second year of implementation, START I far outstripped expectations regarding the numbers of weapons dismantled and the use and efficacy of the inspection regime. By late November, the last Soviet-era nuclear weapon was transported from Belarus and redeployed in Russia. As a result, all nuclear weapons have now been removed from the territories of Belarus, Kazakstan and Ukraine. In this second year, the US carried out 33 START on-site inspections in CIS states (two in Kazakstan, seven in Ukraine and 24 in Russia), and it received a total of 27 on-site inspections.

Very little progress was made on START II, however. Although it was ratified by the US Senate in January 1996, the Treaty had still not been ratified in the Russian Duma by March 1997. The ratification process had been hampered by concerns over: the terms of the Treaty; the state of Russia's conventional military forces; the costs of implementation; NATO enlargement; ABM Treaty demarcation negotiations; and a lack of political focus (due, in part, to the Russian presidential elections in June and to Russian President Boris Yeltsin's ill-health). Although the Duma had established a special commission on START II ratification in February 1996, held high-level discussions with US officials and invited then US Secretary of Defense William Perry to visit the Duma, Moscow legislators remained unconvinced that the Treaty should be ratified.

Instead, Duma members suggested in late 1996 that START II should be by-passed and that Washington and Moscow should negotiate a START III that would supersede the terms of START II. Such proposals were motivated in part by concerns over perceived warhead inequities created by START II. Because this Treaty bans all land-based intercontinental missiles with multiple, independently targetable warheads, Russia would have to destroy a large percentage of its missile force. The US, with more warheads on submarine-based missiles, would not have the same problem. Consequently, the US would have approximately 3,500 deployed warheads and Russia around 3,000 when START II is fully implemented. The Russians would have the right to build 500 new single-warhead missiles to make up the difference, but Duma members were sceptical that the Yeltsin administration would find the necessary funds to do so. This problem was resolved when US President Bill Clinton and President Yeltsin at their summit in Helsinki on 20–21 March 1997 agreed that under START III, both countries would reduce their warheads to 2,000–2,500 by 2007.

At the same time, many Duma members recognise that there are solid budgetary reasons for ratifying START II. Because Russia has ageing missiles and would have to retire them anyway, its strategic forces may well shrink to START II levels or lower over the next few years with little prospect of resources being earmarked for major new building programmes. If this proves to be the case, legislators may decide that it would be better to ratify the Treaty so that the US is similarly forced to reduce its overall numbers.

Up-dating the Anti-Ballistic Missile Treaty

The 1972 ABM Treaty defines the systems it limits as those used 'to counter strategic ballistic missiles or their elements' in-flight trajectory'. Since theatre missile defence (TMD) systems counter shorter-range theatre ballistic missiles, they are outside the scope of the ABM Treaty. But deciding whether or not TMD systems could be adapted for strategic use is far from easy and has always bedevilled interpreters of the Treaty. Washington and Moscow have been discussing the difference between tactical and strategic systems since 1993 in an attempt to clarify this problem. Demarcation between the two types of systems depends on a number of factors, including the velocity of the interceptors and their range.

In November 1993, the US proposed a clarification of the Treaty which would define missile-defence systems according to their targets. An interceptor would be defined as 'strategic' if it could shoot down a target moving faster than 5 kilometres per second (km/sec), which is less than the re-entry speed of an inter-continental ballistic missile (ICBM). Any interceptor not tested against targets moving faster than 5 km/sec would not be counted within the Treaty's limits. Russia objected to this criterion, considering it too permissive and open to the possibility that a strategic missile-defence capability could be constructed by disguising it as a TMD system. Moscow proposed that a defence system should be subject to the ABM Treaty if it had a velocity greater than 3 km/sec, or if it had been tested against a target moving faster than 5 km/sec.

In early 1996, Washington and Moscow finally agreed to settle this issue by splitting the draft agreements into two phases. Phase One would deal with lower-velocity systems, while higher-velocity systems would be left to a second phase. By early October, agreement had been reached on a Phase One agreement, but the US delegation's trip both to the signing ceremony and the Standing Consultative Commission in Geneva in late October was suddenly cancelled. According to Washington, the cancellation was a result of Russia's desire to link the two agreements so that Phase One could not take effect until Phase Two was agreed. While the US had believed that the agreements would be in two distinct parts, Russia was concerned that allowing the first phase to take effect without the second might have left the ABM Treaty vulnerable. At the March 1997 Helsinki summit, the two sides restored momentum to the negotiations by agreeing on elements to demarcate the higher-velocity systems.

Multilateral Regional Arms Control

Adjusting the Conventional Armed Forces in Europe Treaty

The first review conference for the CFE Treaty took place in May 1996 against a backdrop of concern over its future relevance. The long-standing debate over Russia's flank-zone equipment limits was finally resolved at this meeting. Azerbaijan, which had been concerned about Russian military aid to Armenia, removed its objection, allowing the deal to go forward. Russia now has until May 1999 to comply with the new arrangements. The size of the areas bound by the original flank limits (1,300 tanks, 1,380 armoured combat vehicles (ACVs) and 1,680 artillery pieces) has been shrunk to exclude the Pskov, Volgograd, Astrakhan and eastern parts of the Rostov oblasts plus the repair facility at Kushchevskaya and a corridor leading to it. Limits in this new region will be 800 tanks, 3,700 ACVs and 2,400 artillery pieces. In exchange for these alterations, Russia is to receive more inspections in the old and new flank regions and agree to freeze existing flank deployments until the new limits come into effect. There was also agreement that Ukraine could remove the Odessa Military District from the flank region.

The review conference set 15 December 1996 as the date by which all 30 CFE parties must accept the new levels in order for the revisions to enter into force. However, when the Organisation for Security and Cooperation in Europe (OSCE)'s summit convened in Lisbon in December, only 12 parties had notified their acceptance or ratification; the deadline was then extended until 15 May 1997. Even this later deadline may have to be extended yet again: the US Senate is unlikely to vote on the package by that time; Belarus has long ceased providing information for the annual data exchange and has not complied with its national limits; and the political situation in a number of the member-states makes it difficult for them to ratify.

In addition to the new flank-limit agreements, the 30 CFE signatories have begun negotiations to adapt the Treaty to reflect the current political situation in Europe. Ever since the Treaty was signed in November 1990, it has been clear that its bloc-to-bloc structure jarred with political reality. Because the situation had changed dramatically – most notably with the collapse of the Warsaw Treaty Organisation (WTO) in 1991 – it was agreed at the review conference that multilateral negotiations would begin after the Lisbon summit to:

- remove the bloc limits and leave only national limits;
- find a way in which other OSCE states could join the Treaty;
- add new treaty-limited-equipment (TLE) categories, such as transport and airborne warning and control system (AWACS) aircraft; and
- agree lower TLE ceilings.

Plans for NATO expansion played a significant role in Russia's attitude to the CFE Treaty in 1996. Given that the Treaty was originally an agreement that equally affected NATO and the WTO, expanding the Western Alliance has serious implications for the Treaty's meaning. Since the collapse of the Warsaw Pact, Russia has been in a difficult position: while many of its former allies have leaned towards NATO, their equipment is still counted under the CFE Treaty as being within the previously opposing bloc.

Faced with NATO expansion, Moscow has threatened ultimately to withdraw from the CFE Treaty or, at best, to reconsider its commitments. NATO's responded at the end of 1996 by re-examining the Treaty's group structure and reconsidering Alliance limits, including further arms reductions, while trying to avoid a wholesale renegotiation of the agreement.

On 20 February 1997, NATO unveiled radical proposals for revising the CFE Treaty which it had presented to Russia. The plans envisage replacing the bloc-to-bloc and zonal limits with national limits. Signatories would be allowed a limited number of foreign troops on their territory, known as 'stationed forces', but the total of a state's own holdings plus any holdings of the stationed forces could not exceed the national limit. Thus, if Poland or the Czech Republic maintained their current limits they would be unable to station significant amounts of TLE from NATO countries on their territory. There were concerns, however, about whether the CFE agreement could be adjusted in time for NATO's summit in Madrid on 8–9 July. This issue played a major role in discussions between Presidents Clinton and Yeltsin in Helsinki in March.

Controlling Arms in Bosnia

Under the auspices of an OSCE sub-regional forum, substantial progress on arms control in the former Yugoslavia was made in 1996. In January, confidence- and security-building measures (CSBMs) and disarmament talks began among the parties to the conflicts in the region – the Republic of Bosnia and Herzegovina, the Federation of Bosnia and Herzegovina and the Republika Srpska. By the end of the month, the negotiators were prepared to sign an agreement on CSBMs which included:

- restrictions on the numbers, sizes and locations of military deployments and exercises;
- information exchanges;
- limits on the locations of heavy equipment;
- monitoring weapons-manufacturing capabilities; and
- a verification and inspection regime.

While the agreement – and its implementation – was far from perfect, it was a beginning.

Verification of the agreement began in March. A German-led team inspected facilities in the Federation of Bosnia and Herzegovina, and a French-led group conducted inspections in the Republika Srpska. Similar inspections continued between March and August. In June, the parties to the Dayton Accords signed the Agreement on Sub-Regional Arms Control, which limits the numbers of tanks, ACVs, artillery, aircraft and helicopters in a manner similar to that of the CFE Treaty. Data will be exchanged annually and inspections will be carried out. In addition, excess equipment has to be destroyed within 16 months (in a two-phased process). The Agreement also created a Sub-Regional Consultative Commission (SRCC) to hear complaints and deal with compliance concerns.

Regional Nuclear-Weapon Agreements

The concept of Nuclear-Weapon-Free Zones expanded considerably following the signing of the South-East Asia Nuclear-Weapon-Free Zone (SEANWFZ) Treaty in December 1995. The African Nuclear-Weapon-Free Zone (known as the Pelindaba Treaty) was signed on 11 April 1996 in Cairo by 43 African states. Protocols for the Pelindaba Treaty were signed on the same day by China, France, the UK and the US; and Russia signed in November. Although the five nuclear-weapon states have yet to sign the SEANWFZ protocols, some progress was made. In March 1996, two months after France ended its nuclear testing programme, London, Paris and Washington signed the protocols of the South Pacific Nuclear-Weapon-Free Zone (the 1985 Rarotonga Treaty); China and the then Soviet Union had already signed the protocols in the 1980s.

With the signing of the Pelindaba Treaty, the whole of the southern hemisphere landmass is now covered by nuclear-weapon-free zones. An attempt was made at the UN General Assembly in November to declare the entire hemisphere a nuclear-weapon-free zone. This failed, however, when the nuclear-weapon states expressed concern over the implications for naval rights of passage.

Digging Out the Truth in Iraq

The cat and mouse game between UNSCOM and Iraqi President Saddam Hussein continued in much the same vein throughout 1996 as it had in previous years. In March 1996, the UN Security Council approved the export–import surveillance system that will monitor dual-use imports into Iraq if sanctions are lifted. In May, agreement was also reached on the sale of Iraqi oil. Nevertheless, UNSCOM inspectors were repeatedly refused right of entry to suspect sites or were subject to delayed access to facilities where they wished to carry out an inspection. Typically, after difficult negotiations, Baghdad relented and allowed inspectors access in many instances. The suspicion remained, however, that Iraq used the time gained by the delays to modify or remove incriminating evidence.

After four days of talks, Iraq agreed in June 1996 to give immediate, unconditional and unrestricted access to all sites (and to hold bi-monthly meetings with UNSCOM in Baghdad), but in practice this directive has not been fully realised. UNSCOM believes that Iraq is still concealing stocks of chemical weapons, biological agents and a number of missiles and missile launchers – indeed, missile parts were found buried in Iraq. In addition, both the International Atomic Energy Agency (IAEA) and UNSCOM maintain that Iraq is still hiding information and documents on nuclear, chemical, biological and missile programmes.

Multilateral Global Controls

Winning First Prize

It required some political sleight of hand to make it happen, but the CTBT was finally signed in New York on 24 September 1996. There are not many observers, however, who think that the Treaty will ever fully enter into force. India's intransigent position and the wording of the entry-into-force conditions have created overwhelming obstacles.

As the Conference on Disarmament (CD) began its third session of 1996 at the end of July, it became clear that not only would India not support the CTBT, but it would also block the transmittal of the Treaty text from the CD. India did so on 20 August on the grounds that the Treaty did not contain a commitment to a time-bound disarmament framework; that it was discriminatory; and that the entry-into-force conditions, in New Delhi's view aimed at India, were unacceptable.

Article 14 stipulates that the CTBT shall enter into force 180 days after all the states listed in Annex 2 deposit their instruments of ratification, and no earlier than two years after its opening for signature. States listed in Annex 2 include all the nuclear-weapon states and the three 'threshold' nations, India, Israel and Pakistan. In a highly charged atmosphere, India, along with Iran, exercised its right of veto, and blocked the release of the Treaty text from the CD. Drastic action had to be taken. Australia responded by requesting a plenary meeting of the UN General Assembly's fiftieth session (that began in 1995) to consider and vote on the text of the CTBT.

Australia's resolution was co-sponsored by 127 states, so large a number that any attempt to amend the Treaty text at the UN was clearly doomed to fail – in the event none was attempted. The Treaty was debated in New York on 9–10 September. Despite its flaws and the many misgivings expressed during the debate, the Treaty resolution passed with a majority of 158 to three (with five abstentions and 19 absent states). Following its veto at the CD, Iran was widely expected to vote against the CTBT or to abstain, but instead it voted for the resolution. Explaining this negative vote, India's Ambassador Arundhati Ghose said that it had betrayed the 1993 mandate and that India would not sign the Treaty, 'not now, nor later'.

Because of the wording of Article 14, if India does not join the Treaty – and Pakistan will not participate unless India does – its provisions cannot be fully implemented. The CTBT will be signed, and probably ratified, by a large number of countries, all of which can be expected to adhere to the Treaty provisions. Under the 1969 Vienna Convention on the Law of Treaties, states which have signed a treaty are obliged to refrain from any acts that would defeat its purpose. But, without any inspection mechanism in place, it will be difficult to resolve ambiguous seismic signals or to make decisions within an Executive Council. Hence, the issue of entry-into-force conditions is critical.

Article 14 on entry into force also contains provision for a conference to take place if the 'Treaty has not entered into force three years after the date of the anniversary of its opening for signature'. This meeting would examine the extent to which the conditions for entry into force had been met and 'consider and decide by consensus what measures consistent with international law may be undertaken to accelerate the ratification process in order to facilitate early entry into force'. The process would be repeated annually until the Treaty's eventual entry into force. The meeting has been dubbed the 'hand-wringing conference', because it is envisaged that state representatives will be able to do little else. However, it may be the only possibility for implementing the CTBT. Even if it does not enter into force, there is little doubt that the Treaty reinforces progress towards a global norm against nuclear testing. By the end of 1996, the CTBT had gathered 138 signatures.

The Nuclear Non-Proliferation Treaty and Safeguards

At the NPT Review and Extension Conference on 17 April–12 May 1995, the NPT was indefinitely extended. In 1996, prior to the first preparatory committee meeting of the enhanced NPT review process in New York scheduled for 7–8 April 1997, NPT members focused on pursuing a CTBT; establishing a convention banning the production of fissile materials for military purposes; and enhancing IAEA safeguards. In addition, a number of states also concentrated on attempting to start negotiations for a nuclear-weapon convention.

In response to the clear failures of the safeguards system in Iraq, an investigation was commissioned by the IAEA in 1993, known as the '93+2' study because it lasted for two years. The IAEA has been steadily improving nuclear-safeguard initiatives in order to increase its ability to detect clandestine nuclear-weapon programmes. The 93+2 study was divided into two sections. Part One deals with measures already authorised by existing full-scope safeguard agreements. Part Two initiatives require an additional protocol to those safeguard agreements.

Part One provisions began to be implemented from mid-1995. They include such improvements as:

- expanding a state's declaration to include a history of nuclear activities prior to verification of its initial report;
- access to past accounting and operating records;
- details of all activities involving nuclear materials (including research and development);
- no-notice inspections at strategic points; and
- the use of environmental monitoring techniques at those sites.

Part Two measures are more contentious and include:

- declarations on activities relevant to nuclear programmes;
- details of the domestic manufacture of nuclear equipment and materials;
- access for inspectors beyond strategic points at sites and to nuclear-related facilities;
- environmental monitoring anywhere in the state;
- visa-waiver or multi-visa travel arrangements for inspectors; and
- simplified inspector designation procedures.

By early 1996, it became painfully apparent that the passage of Part Two measures was not going to be smooth. Certain states – most notably Germany and Japan – expressed concern over the requirement for information on, and access to, facilities where no nuclear material is held, and particularly the need to protect commercially sensitive information. There has also been much anxiety about the increased reporting and inspection burden on states such as Germany and Japan, which, because of their large nuclear industries are already subject to a great number of inspections. It was also pointed out that the nuclear-weapon states, because they are not subjected to such safeguards, will have the commercial advantage of fewer reporting requirements and inspections.

In mid-1996, a special Safeguards Committee was established to negotiate the Part Two measures. The Committee met in July and October, and by January 1997 had made substantial progress towards agreeing the scope and implementation of the measures. It is hoped that in April 1997 the Special Committee will be able to produce a final text for the additional protocol which could then go to the IAEA Board of Governors in May for approval.

Debating Nuclear Disarmament

Following the renewed commitment to eliminating nuclear weapons expressed at the 1995 NPT Review and Extension Conference, there were some serious attempts to address the question in 1996. In July, the International Court of Justice (ICJ) announced its advisory opinion on the legality of nuclear weapons. The Court had heard evidence from 22 states and the ruling contained something for everyone. The ICJ found that the threat or use of nuclear weapons would 'generally be contrary' to international law, but there was no definite conclusion as to whether the threat or use of nuclear weapons was legal in an extreme case of self-defence 'in which the survival of the state would be a stake'. The Court did conclude that a commitment existed to pursue and finalise negotiations leading to nuclear disarmament: 'the obligation involved here is an obligation to achieve a precise result – nuclear disarmament in all its aspects – by adopting a particular course of conduct, namely, the pursuit of negotiations in good faith'. Everyone claimed victory.

In August 1996, the Canberra Commission on the Elimination of Nuclear Weapons produced its report. The Commission, which included leading military and political figures, was formed by the Australian government in late 1995 in response to decisions made at the NPT Review and Extension Conference. The report concluded that the threats from nuclear proliferation and terrorism are growing and that 'immediate and determined efforts need to be made to rid the world of nuclear weapons and the threat they pose to it'. The Commission proposed a series of 'practical, realistic and mutually reinforcing steps' that could be taken immediately, including:

- taking nuclear forces off alert;
- removing warheads from delivery vehicles;
- ending the deployment of non-strategic weapons;
- conducting further negotiations to reduce US and Russian nuclear arsenals; and
- undertakings for no-first-use and no-use against non-nuclear-weapon states.

In December 1996, one member of the Commission, US General Lee Butler, went further, and in a speech to the US National Press Club called for the abolition of all nuclear weapons. His stance was supported by other generals and admirals from around the world, including General Sir Hugh Beach, Field-Marshal Lord Carver, General John Galvin, General Andrew Goodpaster, General Boris Gromov, General Alexander Lebed and Admiral Antoine Sanguinetti.

Moves to establish an Ad Hoc Committee on Nuclear Weapons, however, were still far from meeting their objective. Although India's linkage of nuclear disarmament and the CTBT had caused much consternation at the CD, the connection between establishing the Ad Hoc Committee and negotiations for a fissile-material convention is much stronger. In early 1997, there appeared to be little room to separate the two issues and remove the current impasse. Consequently, given the strong statements from major nuclear-weapon states opposed to establishing a committee on nuclear weapons, it would seem that a global fissile-material convention is unlikely to be achieved in the near future.

Ratifying the Chemical Weapons Convention

The 1992 CWC reached its target of 65 ratifications in October 1996, allowing it to enter into force on 29 April 1997. Yet the failure of the US Senate and the Russian Duma to ratify the Convention is a severe setback. Although by the end of 1996 there were 161 signatories and 68 parliamentary endorsements, without US and Russian ratification before entry into force, the Treaty will be much impoverished. In addition, the US–

Russia Bilateral Destruction Agreement was seriously delayed because of a lack of money for Russia's destruction programme. The US is continuing to destroy its chemical-weapons stockpile unilaterally.

Despite the US Senate Foreign Relations Committee (SFRC)'s vote in March 1996 in favour of CWC ratification, the Clinton administration decided to postpone the Senate vote which had been scheduled for September. Democratic Senators who favoured the Treaty and were shepherding it through Congress were concerned that opponents had been gaining support, and they felt it was unlikely to win Senate approval. The case against the CWC in the US is somewhat contradictory. On the one hand, some argue that it is unverifiable and would not include some countries of concern. Others claim that the verification regime is too intrusive and will affect US business interests.

Despite the fact that the Convention had been negotiated by Republican administrations and signed by former US President George Bush in January 1993, the arguments against the CWC had clearly become a partisan issue. Republican Senator and SFRC Chairman Jesse Helms stated several times in early 1997 that he would not release the CWC from the Committee unless several of his concerns were addressed. During a speech in March, Helms demanded that key elements be renegotiated, including the verification provisions. He also linked SFRC approval to non-Treaty-specific matters of concern to him, including the abolition of the Arms Control and Disarmament Agency, the US Agency for International Development (AID) and the US Information Agency (USIA). Helms argued that their functions should be incorporated into the Department of State.

Although Clinton and Yeltsin agreed at their Helsinki summit in March to press for CWC ratification, neither country may ratify the CWC before its entry into force. They will thus risk losing the benefits of early ratification, such as a seat on the Executive Council of the Organisation for the Prohibition of Chemical Weapons (OPCW) and staff positions under the treaty implementation programme. This will in turn deal a severe blow to the CWC, which will enter into force without the two states with the largest chemical-weapon stocks in the world.

Verifying the Biological and Toxin Weapons Convention

The 1972 BTWC had run into trouble by the end of 1996. The review conference in November and December made very little progress on the issues of verification and compliance. To a large extent, the lack of advancement was deliberate. The Ad Hoc Group on verification measures and other initiatives to strengthen the Convention, established in 1994, had not been able to complete the draft proposals prior to the review conference. It was decided that it would be better to let the Group complete its work before introducing a protocol. The push (mainly from West

European states) to achieve a verification protocol by 1998 was not part of the final declaration. Rather, the Ad Hoc Group was urged to intensify its work 'with a view to completing it as soon as possible before the fifth review conference', due to be held in 2001.

Confidence in the BTWC has suffered some setbacks over the last few years. In particular, the discovery of the extent of Iraq's biological-weapons programme and the existence of continued gaps in the picture, coupled with concerns over Russia's activities in this field, have led to serious misgivings about the efficacy of the Convention. UNSCOM's experience in Iraq has made it clear that tracking biological and toxin weapons is a painstaking and difficult process, and complete confidence in a verification regime may be impossible. It is important, therefore, to be realistic when assessing the Convention and its effect. Once a verification structure is in place, confidence may grow. But, because of the difficulty of tracking such materials and activities, particularly with rapid advances in biotechnology, it is important not to be lulled into a false sense of security. To strengthen the verification regime, and to try to ensure compliance with the BTWC, other deterrent measures, such as adequate defences, are therefore essential.

Prohibiting the Use of Inhumane Weapons

After an inconclusive review conference of the 1981 Convention on Prohibitions or Restrictions of the Use of Certain Conventional Weapons (CCW or Inhumane Weapons Convention) in September and October 1995, the conference reconvened in January, April and May 1996 to continue considering a ban or limitations on land-mine production and deployment. By the beginning of May, it had reached consensus on restricting the use of anti-personnel land-mines. Mines not located within a mapped and guarded area will have to possess a device capable of neutralising it within 120 days of being laid. All mines with a metal mass of no less than 8g must be detectable. The parties have nine years to bring existing stocks in to line with the new regulations.

Throughout the rest of 1996, support for a global ban on land-mines came from development and humanitarian-aid agencies, military generals, and even the Pope and the UK's Princess of Wales. In October, Canada established the Ottawa Conference to hasten a global ban on anti-personnel mines. Only China seemed to be prepared to oppose the global trend publicly.

A Very Mixed Picture

Despite the potential problems for the CTBT, 1996 will be remembered as the year in which nuclear testing was eliminated. If the Treaty manages to establish an international norm against nuclear testing, China will have conducted the last ever nuclear-weapon test on 29 July 1996. The precise implications of the CTBT for nuclear-weapon programmes have yet to be fully understood, but it is likely that the Treaty will have a profound effect on the future programmes of nuclear-weapon states.

The NATO expansion debate also had a prominent effect on many negotiations in 1996, in particular, the START II process, the ABM Treaty and the CFE Treaty. In some respects, the drive to adapt the CFE Treaty may well become one of the most constructive outcomes of the NATO expansion debate. Restructuring the Treaty to take into account political change in Europe is a positive development.

In other areas, slow – or no – progress has left much to be concerned about. The demarcation between theatre and strategic missile defences must be clarified. China is nervously watching this process, as it may affect the prospects for missile defences in North-east Asia. The CWC has run into trouble because of resistance from both Chairman of the US SFRC Helms and the Russian Duma. A CWC without the US and Russia, two of the initial states parties, will be an impoverished Convention; and a CWC without the participation of the US and Russia at all would be a disaster.

The NPT enhanced review mechanism begins in 1997. Following the CTBT, the US and Russia are positioning themselves to begin START III negotiations. The nuclear-weapon states are likely to find that pressure to open nuclear disarmament negotiations will mount over the next few years. There is a growing group of countries, including many allies of the nuclear-weapon states, which would like to see such progress, and this pressure may gradually become harder to resist.

The Americas

In November 1996, US President Bill Clinton became the first Democrat since Franklin Roosevelt more than half a century earlier to be re-elected for a second term. Although he swept to an easy victory, his hopes for full support were frustrated by the return of Republican majorities to both houses of Congress. The result guaranteed bipartisan bickering that will seriously restrict White House initiatives in both domestic and foreign affairs. Even before the election, Clinton was constrained by the conservative tendency of both the public at large and Congress; during 1996, the impact of domestic considerations loomed larger in the formation of foreign policy than has been the case for many years.

The difficulties facing the President became clear almost immediately by the spread of a new scandal concerning the funding of the Democratic election campaign. Accusations that illegal contributions had been gathered from abroad were denied by the administration, but the campaign office returned hundreds of thousands of dollars from questionable sources. Even Vice-President Al Gore, who had been the clear choice as Democratic candidate for the next election in 2000, was tarred by his over-zealous collection activities.

The Clinton administration's dealings with its Latin American neighbours epitomised the dilemma produced by the lack of a clear mandate. Clinton would probably like to adjust out-of-date US policies towards Cuba, but finds this impossible in view of the Republican's hard line against Cuban President Fidel Castro. Washington's heavy-handed 'war on drugs', another policy that many in the administration feel could be fine-tuned, is also hostage to the conservative voices in Congress. Yet it is creating new tensions in Washington's relationship with many Latin American countries, while contributing to the growing problem of deteriorating civil–military relations in some of these countries as well. On the whole, Latin America struggled with issues of security, democracy and economic growth, avoiding stumbling into the many existing pitfalls, but not surging forward in their efforts to consolidate overall gains in these areas.

The United States: Basking in Security

As is always the case in a US election year, the intense focus on the Presidential and Congressional campaigns ensured that foreign and defence policy were not high on the administration's agenda in 1996. Since

President Bill Clinton was far ahead of his rival Bob Dole in the polls from the beginning of the year, focusing on the election placed a premium on doing nothing wrong. In his quest for the centre ground, Clinton discovered a majority satisfied with the status quo and distrustful of governmental policy experiments, at home or abroad. While some decried what they saw as inaction and lack of vision, most of the electorate appeared to cherish the unprecedented security, prosperity and stability they believe they have earned through hard work and a firm stance against communism.

Clinton's attempt early in his first term to define US foreign policy as a new, grand project to enlarge the global democratic sphere had earned him tepid public support. In contrast, during his campaigning in 1996, Clinton recognised and played to his people's foreign-policy conservatism. Paradoxically, the administration also realised that the power of today's media means that large-scale human suffering, almost anywhere in the world, cannot go unaddressed, but that doing so chips away at the electorate's sense of absolute security. Clinton dealt with this paradox by avoiding grand commitments and strategic visions, but also by being willing to commit resources, on a case-by-case basis, to problem areas such as Bosnia and Haiti where US power could cheaply (especially in terms of lives) lead the way in resolving the problem.

In a time of great international uncertainty, it is probably wiser to risk being attacked for lacking strategic vision than to set forth a policy which falsely claims to have identified and prepared for future threats. Yet, in the dynamic world of international politics, maintaining the status quo requires well-conceived, purposeful action. In the long term, making sure that nothing threatening happens is quite different from doing nothing. The administration's *ad hoc* responses in the Middle East, towards China and Taiwan, and at the UN during the last two years of Clinton's first administration carefully threaded a way through this conundrum. Although in each case some success was attained, there were also setbacks.

Clinton's challenge in his second term is to sustain interest and attention in those areas where US actions have already made a positive difference, but where further action is still required to ensure continued success. Clinton's new Secretary of State, Madeleine Albright, began well in February 1997 on her first tour to areas where US initiatives will continue to be important – Europe, the Near East, Russia and the Far East, particularly China. Each region presents difficult problems that require adroit handling to prevent them worsening. Each, however, also shows some hope of resolution that would probably be lost without active US efforts.

The End of 'Big Government'

After the bitterly partisan battles that marked the Republican Party's attempt in 1995 to implement the 'Contract with America', both Republicans and Democrats appeared to conclude in 1996 that the public was weary of unseemly bickering and government shutdowns. At the end of 1995, the Republican-led Congress had thought that by crippling government functions by refusing to pass the necessary enabling legislation, Clinton would buckle. The President proved to be a better judge of the public mood, however, and he essentially won the budget stand-off in January 1996 when public frustration with the shutdown finally forced Congress to back down. The subsequent sharp swing in the opinion polls away from Leader of the House Newt Gingrich's revolution convinced both sides that the public wished to preserve the status quo, even though they complained loudly of its flaws. In response, and in great contrast to 1995, few major domestic initiatives were attempted in 1996 and even fewer attracted much attention. Instead, attention concentrated on the Presidential and Congressional election campaign.

With no challenge from within his own Democratic Party, Clinton was able to avoid the kind of damaging primary campaign that had so hurt George Bush in 1992. Instead, he sat regally on the sidelines while the Republicans battered each other in the primaries. Senator Bob Dole survived early setbacks in New Hampshire, Delaware and Arizona and, in the end, easily secured the Republican nomination. But in the process he was forced by right-wing social conservatives and supply-siders to move away from more centrist positions on abortion, deficit reduction and tobacco regulation which might have appealed to the wider electorate in November. Clinton seized that centre ground by espousing the more moderate Republican positions that Bob Dole had been forced to renounce. Even though Dole attempted to retake this ground after the primaries, he found it firmly occupied by the President.

Compared to previous elections, the campaign itself was muted as the candidates moved ever closer on issues of real significance. This ideological similarity, combined with a robust economy and lack of foreign-policy disasters, provided little ammunition for Dole. The public quickly grew bored with repetitive attacks on the President's character, and the lack of any substantive argument created deep voter apathy. The result was the re-election of the President on 5 November 1996 with the lowest voter turnout – 49% – in US presidential election history. As if to reinforce the President's lack of mandate, the public also returned a Republican majority in both houses of Congress. The pre-election totals were 236 Republicans to 198 Democrats, with one independent, in the House of Representatives, and 53 Republicans to 47 Democrats in the Senate. The election changed these figures to 227 and 207, with one independent, in the House, and 55 and 45 in the Senate.

The President's election campaign combined ideological flexibility with an acute sense of how the average voter felt. When the Republican majority squandered the initiative it had gained in the 1994 elections, Clinton did not repeat his rivals' mistake of confusing temporary popularity in the polls with a mandate for change. Instead, he adopted an incrementalist programme, emphasising mostly low-cost, low-impact domestic initiatives such as promoting technology to protect children from violent television images, a bill to allow family members leave from work when another member is ill, and others to encourage the use of school uniforms and to ban gay marriage. At the same time, he finally shook off the Democrats' image of fiscal irresponsibility by declaring in his 1996 State of the Union address that 'the era of big government is over', and by matching the Republican's plan to balance the budget by 2002. Together, these measures projected the image of a President sensitive to the economic and family issues that trouble middle-class Americans (especially women), but one who also championed fiscal responsibility and had a healthy distrust of purely governmental solutions to social ills. Politically, they enabled the President to co-opt the Republican mantle of fiscal rectitude without also inheriting their image of callousness to the needs of the disadvantaged.

The principal exception to this trend of non-activism was the Welfare Reform Bill, which Clinton reluctantly signed into law on 22 August 1996. During his 1992 campaign, Clinton had promised to 'end welfare as we know it'. The Republican-prepared Bill represented a truly profound reshaping of the US welfare system by devolving much of the decision-making power to the individual states and thus, for the first time, ending a Federal entitlement. Clinton expressed major reservations about the Bill, especially over provisions that denied support for legal immigrants, but he had already vetoed two Republican bills that even more radically changed the welfare system. For the sake of his re-election chances, he had to concede this third attempt. When he signed the Bill into law, he promised to rectify its flaws after the election.

During 1996, the Democrats, often led by Senator Edward Kennedy, managed to cajole the Republican Congress into raising the minimum wage from $4.25 to $5.15 an hour and approving a bill allowing workers to carry their health insurance with them when they leave or lose their jobs. One other major piece of legislation concerned internal security. On 9 October, in reaction to the spate of possible terrorist acts in the United States – including the explosion of TWA flight 800 on 17 July and the Atlanta centennial park bombing on 27 July – Congress passed, and the President signed, tough anti-terrorism legislation enhancing airport security and giving the law-enforcement agencies new powers to combat terrorism on US soil.

While these measures were undoubtedly important and, in the case of welfare reform, possibly even transforming, they contributed little to continuing efforts to balance the budget (welfare reform will save only $54.6 billion to 2002). Welfare reform affects only the politically marginal young poor, while the programmes with the greatest budgetary impact, Social Security and Medicare/Medicaid, directly concern middle-class and especially elderly interests. The conservatism of these politically potent interest groups meant that Social Security and Medicare/Medicaid remained largely unreformed in Clinton's plan to balance the federal budget. These three entitlement programmes, with 38% of the federal budget in fiscal year 1995, already dwarf all other entitlement programmes, including welfare (6%) and even defence (18%). Their huge projected growth implies that, even if the budget can be balanced by 2002 under the Clinton plan, the hard decisions necessary to eliminate the structural budget deficit beyond that date have not yet been taken.

Foreign Policy: Unilaterally for Peace

Developments in 1996 completed the 'on-the-job' foreign-policy training that the Clinton administration had been engaged in since it took office. If the first two years were characterised by a relative indifference to foreign affairs and the third year by a heady reassertion of US leadership, 1996 reinforced the message that the outside world is a dangerous place for US political leaders. In 1995, partly as a result of Clinton's need to demonstrate his relevance after the Republicans seized the domestic agenda, the US leadership had concentrated on foreign problems, contributing significantly to the search for peace in such areas as Northern Ireland, Bosnia, Haiti and the Middle East.

Clinton basked in the glow of these mediation efforts, which ended with a triumphant voyage to Northern Ireland in December 1995. Within a year, however, the potential downside of foreign-policy activism was revealed. The fragility of the agreements in Bosnia, Northern Ireland, Korea, and the Middle East became clear. There were no conspicuous disasters, but on several fronts they were just barely avoided.

Nowhere was this more evident than in the Middle East. The moment that defined Clinton's public persona as a statesman and the US as a world leader dedicated to peace was the historic September 1993 handshake on the White House lawn between Israeli Prime Minister Yitzhak Rabin and Palestinian Chairman Yasser Arafat that launched the formal Israeli–Palestinian peace process. In 1996, however, the defeat of Israel's Labour party candidate Shimon Peres (whom Clinton had overtly supported and championed) by the *Likud*'s Binyamin Netanyahu in elections on 30 May further threatened a peace process already reeling from several random terrorist attacks against Israelis and a deadly Israeli incursion into Lebanon the previous April.

It took six months of concerted effort for Netanyahu, who had rejected the idea of ever meeting Arafat, and Arafat, who was beginning to threaten to return to street actions, to agree on how to implement the Oslo peace accords. Throughout this period the US retained its role as the mediator of choice, providing the necessary impetus for new advances. Clinton convened an emergency summit at the White House on 2 October, attended by Arafat, Netanyahu and King Hussein of Jordan, and after impassioned pleas from King Hussein and Clinton the two sides began to move towards an agreement. On 22 December, Clinton also dispatched envoy Dennis Ross to the Middle East and, after long negotiations, agreement on how to continue the peace process was reached on 14 January 1997.

In the Persian Gulf, however, the US presence appeared significantly less benign, at least to some of the parties. Apparently in protest at the US military presence in Saudi Arabia, a US military barracks in Dhahran was bombed on 25 June, killing 19 servicemen. Amid claims of lax security, 4,500 US troops were moved to a more isolated location, although still within Saudi Arabia. Saudi authorities have blamed Iran for the attack, but have restricted the Federal Bureau of Investigation (FBI)'s ability to conduct a full investigation, fuelling speculation that the attack was carried out by domestic groups opposed to the monarchy's Western alliances. Elsewhere in the Gulf, Iraq's ever-resilient Saddam Hussein also challenged US power when, in late August, he sent 40,000 troops to intervene in an intra-Kurdish civil war in the autonomous Kurdish enclave in Northern Iraq. The US responded on 3 September by launching 44 *Tomahawk* cruise missiles against Iraqi air-defence sites in southern Iraq and by extending the no-fly zone imposed after the 1991 Gulf War almost to the suburbs of Baghdad. While domestically this tough action may have helped dismiss Dole's claims that Clinton's foreign policy was feckless, the failure to win UN, or even most allies' approval for the attack, struck many as a unilateral abuse of power. As a result, France refused to help patrol the extended no-fly zone.

Clinton's greatest foreign-policy gamble was in the former Yugoslavia. In December 1995, when he sent 20,000 US troops to Bosnia to help enforce the General Framework Agreement on Peace in Bosnia and Herzegovina, reached in Dayton, Ohio, he did so against considerable opposition. Polls indicated that 55% of the US public disapproved of the action at the time. In the event, the gamble worked, at least in the short term. The NATO-led Implementation Force (IFOR) that took over from the UN Protection Force (UNPROFOR) in December 1995 largely succeeded in separating the combatants and confiscating their heavy weapons. It took no casualties from hostile action and Bosnia remained a peaceful, if not harmonious, place throughout 1996. Compared to the ethnic cleansing horrors of 1995, these were substantial accomplishments. From an electoral perspective, Bosnia also stayed out of the news, thus reassuring the US public that the world was a stable place and denying Bob Dole the principal foreign-policy stick with which candidate Bill Clinton had beaten George Bush in 1992.

Although President Clinton had originally insisted that US troops would be withdrawn from the Balkans after a year, he quickly recognised that a durable peace would not be achieved by then and that without an international peacekeeping force in place, renewed fighting would be likely. Thus, on 15 November, he agreed to a 31,000-member, 18-month follow-on force to IFOR, the Stabilisation Force (SFOR), including 8,500 US troops.

The agreement to maintain this Force masks the divide in the NATO alliance over how to bring lasting peace to the area. The US wants to build up the Bosnian Muslims and the Croat Federation militarily, while isolating the Serbs as a punishment for aggression and atrocities. The Europeans favour concentrating on regional arms control and economic reconstruction. This disagreement gives hope to all parties to the conflict, discourages compromise and ensures that the peace will remain a fragile one for some time.

IFOR's success has masked these disagreements over Bosnia from the public. However, the US and its allies were less successful elsewhere, as intra-Alliance disagreements spilled over into the headlines in 1996. One conspicuous example was the dispute over the reappointment of Boutros Boutros-Ghali as UN Secretary-General at the end of 1996. The US held tenaciously to the line that Boutros-Ghali had failed to reform the wasteful UN bureaucracy and should step down. Both US adversaries and allies, particularly France, disagreed. In the end, on 19 November, the single US veto outweighed the otherwise unanimous support of the 15-member Security Council and Boutros-Ghali failed in his bid for re-election. Instead, the Security Council elected Ghana's Kofi Annan, after France eventually withdrew its opposition.

Another major source of inter-allied dispute erupted over US relations with Cuba. On 24 February 1996, Cuban warplanes shot down, either in or near Cuban airspace, two small US-registered civilian aircraft owned by a Cuban exile organisation. Spurred by that provocation and pushed by the politically powerful Cuban expatriate community in Florida, Clinton signed the Cuban Liberty and Democratic Solidarity Act (better known as the Helms–Burton Act) on 12 March which tightened the US embargo against Cuba. The Act's most controversial provision allows American citizens to seek compensation in US courts from foreign companies that use formerly US-owned property expropriated by Cuba. It also threatened to bar the executives of such foreign companies from entering the US. The President signed a similar bill on 5 August, known as the D'Amato Act, which threatened sanctions against foreign companies which invested heavily in Iran or Libya. Canada and the European Union, particularly, reacted to these novel extra-territorial applications of US law with indignation and threats of retaliation. Clinton judiciously delayed full implementation of the Helms–Burton Act until mid-1997 and, in the meantime, is attempting to reach some *modus vivendi* with Canada and the EU.

The NATO alliance did agree, at its December 1996 ministerial meeting in Brussels, to hold a special summit on 8–9 July 1997 in Madrid to invite new candidates to begin negotiations to join the organisation, but did not specify which might be admitted. The most likely candidates are Poland, the Czech Republic and Hungary, but a number of other countries still hope to be considered. The main impediment is Russia's attitude towards expansion. Russian President Boris Yeltsin's government expresses almost daily its staunch opposition to the concept, as forcefully reiterated by Foreign Minister Yevgeny Primakov at the December ministerial meeting. However, Moscow's willingness to cooperate with NATO over other issues, and its occasional inconsistencies regarding enlargement, have led some US officials to speculate that rhetorical Russian opposition to expansion is only a bargaining position. Divining the true Russian position is certainly difficult given the tumultuous state of its domestic politics.

In Asia, the world's other potential superpower, China, has made abundantly clear its suspicion of the US. Sino-US relations were in a dire state at the start of 1996, but during the course of the year they improved significantly. The usual wrangle of complaints and counter-complaints over trade, human rights and missile sales to Pakistan and Iran were exacerbated by China's actions towards Taiwan. In an attempt to influence the first ever Taiwanese presidential elections on 23 March 1996, China conducted large-scale military exercises across the Taiwan Strait and carried out missile tests in the water close to two of Taiwan's busiest ports. The US responded by sending two aircraft carrier battle groups to the area. The crisis passed after the Taiwanese elections, only for tensions to rise again when on 15 May the US accused China of violating a February 1995 agreement to reduce Chinese piracy of US intellectual property. The US threatened trade sanctions against $3bn-worth of Chinese exports if China did not fully honour the agreement. China immediately threatened retaliation against US goods.

Once that difficult beginning was over, however, relations between the US and China improved dramatically. By 17 June, after several meetings between US Trade Representative Charlene Barshetsky and Chinese President Jiang Zemin, the US declared that China had taken 'serious and important steps' towards implementing the 1995 agreement and cancelled the sanctions. Also in June, Clinton once again renewed China's most-favoured-nation (MFN) trading status, despite the administration's own conclusion that human rights in China had deteriorated in the last year. The thaw in US–Chinese relations culminated in a decision announced at the December Asia-Pacific Economic Cooperation (APEC) summit in Manila that Clinton and Jiang Zemin would exchange state visits in 1997, the first such visits between the two countries since the Tiananmen massacre in 1989. The US–Chinese relationship, however, remains troubled with trade, human rights, proliferation and Taiwanese issues bound to cause friction in the near future. The US is still blocking China's entry into the World Trade Organisation (WTO) on the grounds that Chinese protectionist barriers must first be lifted. China, for its part, believes its developing-country status should allow entry to the WTO despite its relatively protected markets.

Washington's growing preoccupation with China has also changed the tenor of its relationship with Japan. In July, for the first time, the US had a greater monthly bilateral trade deficit with China than with Japan. This economic trend, together with growing Chinese military assertiveness, makes the US and Japan less interested in engaging in their own trade disputes than in shoring up their military alliance. Thus, Clinton and Japanese Prime Minister Ryutaro Hashimoto reaffirmed their 1960 Japan–US Security Treaty on 17 April in Tokyo. This 'joint security declaration' pledges the United States to maintain its military commitments in East Asia at their current level of 100,000 troops, including 47,000 in Japan. In return, the Japanese gave assurances that they would provide assistance for military contingencies other than defending Japan itself. China immediately claimed that this increased US–Japanese military cooperation could only be directed at containing Chinese power, since there was no longer any Soviet threat to Japan.

The Clinton administration thus managed to forestall any disasters in 1996 that might have disrupted the President's electoral prospects, but in the process upset many allies and potential adversaries with its renewed tendency to assert unilateral US power and leadership. The United States' special position in the world offers unique opportunities to provide leadership and pursue peace, but also unique opportunities to abuse that position for domestic or international gain. In 1996, US foreign policy took advantage of both opportunities.

Defence Policy

Unlike the brutal budget battles and government shutdowns in 1995, a substantial bipartisan consensus on defence budget issues unfolded in 1996. The funding levels supported by Congress and the Department of Defense, especially over the five-year planning horizon, differ only minimally. Congress even backtracked over the most controversial aspect of last year's bill when, on 24 April, it repealed a law mandating that HIV-infected military personnel be discharged. There are, of course, some substantial disagreements between Congress and the White House, particularly regarding certain high-profile weapons programmes, the timing of force modernisation, and the whole issue of ballistic-missile defence (BMD). But these problems were largely swept under the carpet as both Congress and the President felt the need in an election year to demonstrate their ability to run the government.

On 23 September, Clinton signed the fiscal year (FY) 1997 Defense Authorization Bill, which called for $265.9bn in defence spending, about $11.2bn more than the administration had initially requested. Most of these additional funds were to purchase four *Aegis* cruisers ($3.4bn), F/A-18 and F-22 aircraft ($4.2bn) and additional research-and-development funds for BMD ($0.96bn). Over the course of the Future Years' Defense Plan (FYDP) for FY1998–2002, Congressional and Presidential spending plans differ by less than 1% ($1.36 trillion for the President, $1.37tr for Congress). This year's spending differential resulted from the administration's desire to delay some procurement to the end of the FYDP rather than from a difference about the overall level. Moreover, the Defense Authorization Bill did not deviate from the administration's 1993 'Bottom-Up Review' (BUR) strategy which calls for the US to be able to fight two 'major regional contingencies', or wars, simultaneously. Both requests also signal a levelling out, at about two-thirds of the peak Cold War purchasing-power level, of the post-Cold War decline in defence spending. After almost a decade of reductions, the BUR force level of 1.4 million personnel, 10 army divisions, 11 aircraft carrier groups and 3 marine divisions is now essentially in place.

The relatively painless passage of the 1997 Bill, however, should not obscure the substantial debates still raging over defence policy. The Defense Authorization Bill called for the Department of Defense to conduct in 1997 a quadrennial defence review to completely re-examine the 'defence strategy, force structure, force-modernisation plans, budget plans, and other elements of the defence program'. The defence review will reopen the question of US defence planning and reveal some severe disagreements. The greatest debates will probably focus on the BUR's two-war requirement, ballistic-missile defence, the budget trade-off between maintaining readiness and force modernisation, and US nuclear strategy.

The defence review will need to ask three questions. First and foremost, is a two-war standard appropriate for the US? The BUR requires the US to have the capability rapidly to win two simultaneous, low-casualty, unilateral victories over medium-sized states; the most obvious candidates are North Korea and Iraq. Many budget analysts feel that the two-war standard was created simply to justify the near-Cold War level of spending that has allowed politicians to protect most of the defence industry and armed forces jobs created by the Cold War build-up. They note that this huge force, relative to that of other nations, is based on the implausible assumption that the US would ever wage two large-scale wars alone and do so against small countries of limited strategic import. Others feel that the US needs a larger force to deter potential competitors and, failing that, to provide a base for any future superpower competition. The two-war standard will probably remain, however, if only because no other reasonable force-sizing strategy can sustain the current spending level.

If this standard does remain, the next question is: will the current downsized force meet the two-war standard? Recent humanitarian and peacekeeping operations, such as in Haiti and Bosnia, have led to complaints that these operations have eroded the military's ability to meet the two-war standard. Moreover, current modernisation plans do not provide enough procurement budget to adequately replace and modernise existing equipment. In contrast, some doves claim that the military capabilities of low-tech Iraq and impoverished North Korea have been exaggerated and that a smaller force would be adequate.

The final question is whether the Presidential and Congressional budgets adequately fund this force. Recent General Accounting Office (GAO) and Congressional Budget Office (CBO) reports have placed the FYDP mismatch between the plans and actual funding at $150bn and $47bn respectively. The administration counters, however, that these reports fail to take into account savings from lower inflation, acquisition reform and base closures. Nonetheless, projected shortfalls may allow advocates of the so-called revolution in military affairs to gain acceptance for a more radical force restructuring which claims to be able to deliver equivalent or better combat power at lower cost through synergistic exploitation of new technologies, especially stealth, information technology and precision-guided munitions.

The most acrimonious debate will probably be reserved for the issue of ballistic-missile defence. Republicans in Congress remain adamant that the US should deploy a national missile defence as soon as possible to protect against small-scale ballistic-missile attacks. The administration has provided funding for theatre missile defence systems such as the PAC-3 *Patriot* follow-on and the Navy Lower Tier, but believes that any effective national missile-defence system, and possibly many of the more sophisticated theatre systems, would require the US unilaterally to abrogate the 1972 Anti-Ballistic Missile (ABM) Treaty. This abrogation would, in turn, threaten Russian cooperation on a variety of arms-control issues, particularly the ratification of the Strategic Arms Reduction Talks (START II) Treaty now mired in the Russian Duma.

Opponents of BMD also cite November 1995 National Intelligence Estimate 95-19, 'Emerging Threats to North America During the Next 15 Years', which predicts that there will be no ballistic-missile threat against the continental US for at least 15 years, and a May 1996 CBO estimate that a national system could cost as much as $60bn. The 1997 Defense Authorization Bill included the Defend America Act which funded research for both advanced theatre systems and a national system, but, at Clinton's insistence, did not set a timetable for deployment while the administration attempts to negotiate a reinterpretation of the ABM Treaty with Russia.

Usually unspoken in this debate is the belief of many experts that no effective national missile defence is possible and that such a system would only create the illusion of nullifying the Russian and Chinese arsenals, thus spurring an unnecessary nuclear-arms race. The debate is further complicated by polls showing that the majority of Americans believe that some form of national missile defence already exists. The Americans are thus both unwilling to pay for a 'new' defence and unwilling to accept the consequences of not having a pre-existing one. As a result, neither side is able to generate much popular enthusiasm, but politicians on all sides recoil in fear at the public anger that would result from a deadly missile attack on US soil.

The quadrennial defence review will also include a nuclear strategy review. The idea that the current situation might allow dramatic changes in the US nuclear posture has taken on new prominence since December 1996 when General Lee Butler, former commander of the Strategic Air Command, led a group of 40 retired generals in a call for complete nuclear disarmament. This campaign reflects the view of many that the current situation offers an opportunity to begin dramatic reductions in nuclear weapons. Indeed, Russia is reluctant to ratify START II in part because it will be required to dismantle so many multiple-warhead missiles that it will only be able to reach the Treaty's warhead ceiling by substantially reinvesting in single-warhead missiles – an investment it clearly cannot afford. Therefore, passage of START II may be conditional on US acceptance of a START III that mandates reductions to perhaps 2,000 warheads per side. From the perspective of the military community, nuclear disarmament offers the opportunity to tap into the $33bn per year currently spent on nuclear infrastructure in order to pay for conventional arms modernisation. These conjunctural effects may mean that the nuclear strategy review will contain some dramatic recommendations.

Neither Isolationism nor Dominance

Although Clinton was originally reluctant to engage himself deeply in foreign policy, like most US Presidents he eventually discovered that the foreign arena offered him more flexibility and less politically charged problems than domestic policy. As a consequence, far from leading the US into renewed isolation, Clinton involved it more in world affairs during his first term than many expected. Clinton's experience also showed that, despite occasional isolationist populism in the US, the experience of two generations of world leadership has created a public and political system which thrives on engagement in the world, even as it insists on absolute security at home. The task of Clinton's second term will be to define that engagement and perhaps more importantly that leadership in terms self-interested enough to sustain US public support, and yet soft-spoken enough to reassure the world that its leadership is not a prelude to US dominance.

Security Versus Democracy in Latin America

Security, democracy and economic growth continue to shape politics in Latin America. After the heady days of democratic transitions and market reforms in the 1980s, there is a growing awareness that democracy is still shaky in many countries of the region. New security threats continue to be a major challenge. At the top of the list is drug trafficking, which has given rise to major criminal organisations and has led to an increase in terrorist and violent actions. Although drug syndicates pose an armed threat to the state, their main challenge to governmental authority is through the corrosive influence of corruption, perhaps the greatest obstacle to democratic consolidation in the region.

The rise of the criminal drug trade has also led to a reorganisation of US military and foreign assistance policies in the Americas, some of which are designed to strengthen local judicial systems and attack corruption. The majority of the anti-narcotics programmes, however, are based on military and police assistance. These have replaced the anti-communist security-assistance programmes of the past three decades.

The new policies are as controversial as the old; in some cases, US anti-narcotics schemes have created major tension between Washington and its Latin American allies. Most controversial is the 1986 US law requiring the US President to certify each year those countries that are cooperating in the 'war on drugs'. Until last year, Washington had always certified allied countries, reserving de-certification for those perceived as hostile to US interests. This changed with the de-certification of Colombia in March 1996.

De-certification requires the US to abstain from supporting international loans, to suspend all assistance except for anti-narcotics and humanitarian needs, and to consider economic sanctions. It also requires the US to identify those sectors that are cooperating within the sanctioned country. In Colombia, this has led to stronger ties between the US and the Colombian armed forces and police, while by-passing the President of Colombia and most of the executive branch. The result is a policy that strengthens the military and weakens civilian authority, a policy at odds with the prevailing trend of supporting civilian, democratic governments throughout Latin America. Although the context and times have changed, this approach echoes past mistakes when the US favoured Latin American militaries in the fight against communism and subversion.

Central America and the Caribbean

The Guatemalan civil war ended in 1996. After almost a decade of negotiations, international mediation and partial accords, a peace treaty was finally signed between the *Unidad Revolucionaria Nacional Guatemalteca* (URNG) and the government on 26 December 1996. By the time the conflict ended, both the guerrillas and the government had changed dramatically. The guerrillas, having been pushed deep into Guatemala's interior, emerged as a voice for the nation's majority indigenous populations. Over the course of the arduous negotiations, they won major concessions on human rights, repatriation of refugees and recognition of indigenous cultural identity.

The government, too, had held regular elections and gradually consolidated civilian rule since the 1980s. One of the final agreements between the two sides was designed to strengthen civilian control over the armed forces and intelligence services by reducing the size of the military, reorganising the police and demobilising and disarming civilian defence patrols.

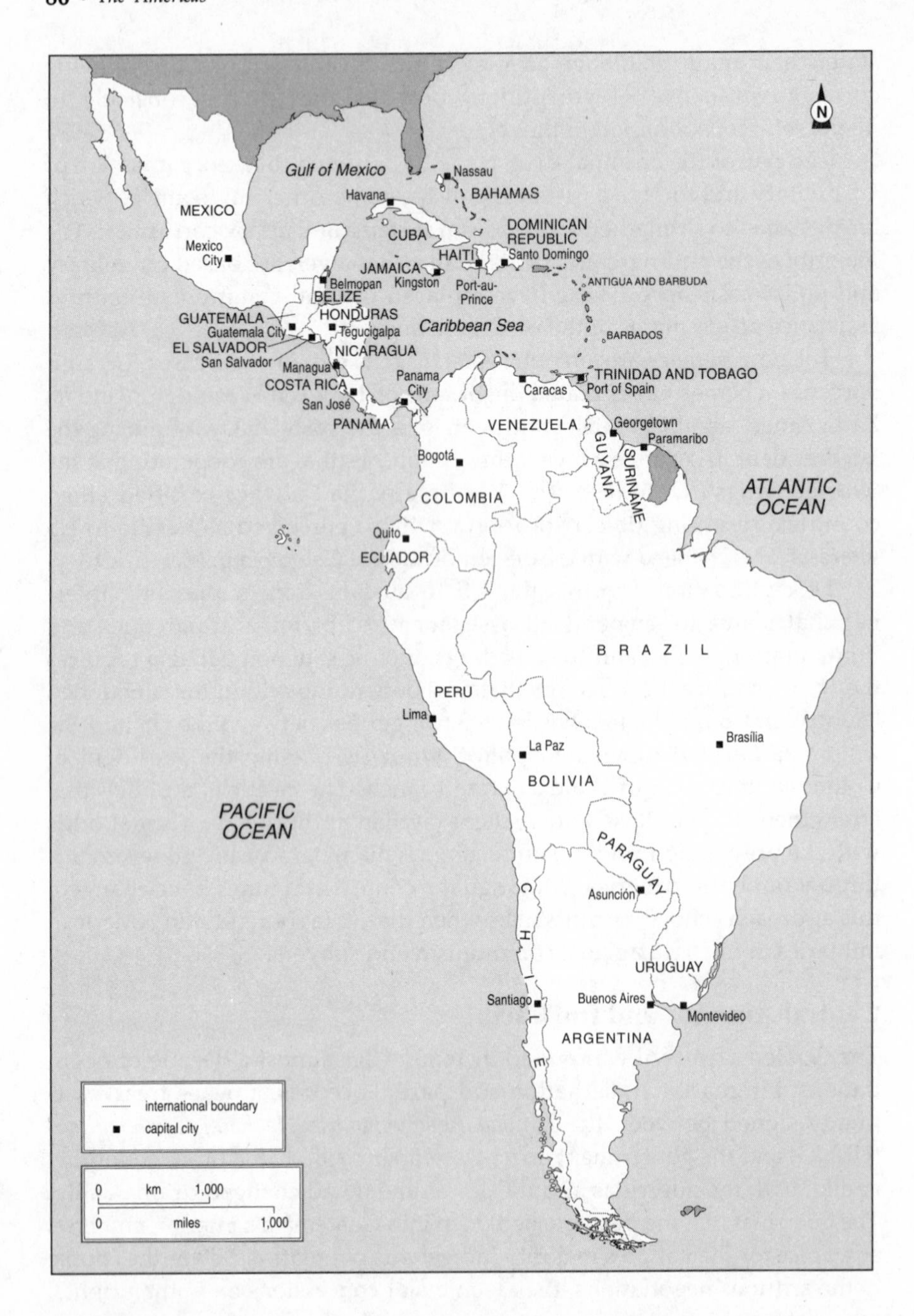

The URNG aspires to become a political party, following the path of several other former guerrilla movements in Latin America. Under the peace accords, crimes and human-rights abuses by both sides will be pardoned, except in cases of genocide, torture and forced 'disappearances'. The Inter-

American Development Bank has approved a $500 million loan to support the peace process and national reconciliation over the next two years.

The end of the civil war in Guatemala completes the Central American peace process. Yet this has not always brought an end to violence. In El Salvador, there has been a major escalation in violent crimes committed by youth gangs and demobilised guerrillas and army officers. Nicaragua, too, has suffered from a rise in social violence, suggesting that the legacy of political brutality may be major increases in the crime rate for years to come.

In addition, Nicaragua finds itself still confronting the legacy of the 1980s insurgencies and civil conflict. It was once widely expected that the Sandinistas, the largest and most organised political party, would easily return to power in 1996, following their unexpected defeat at the polls in 1990. Yet as the November 1996 elections approached, they split into two groups, one led by former Vice-President Sergio Ramírez, and the other, more militant one led by former President Daniel Ortega. Ortega became the Sandinistas' candidate, but despite an impressive advertising campaign that almost succeeded in creating a new political image for him, he lost to the former conservative Mayor of Managua, Arnoldo Alemán Lacayo, of the *Alianza Liberal*, whose party won 42 of the 93 seats in Congress.

The Sandinistas still wield significant power in Congress, the labour unions, the armed forces and police, city halls and elsewhere in society, and Alemán has pledged to work with them. But following the election the two sides quickly fell out, first over the legitimacy and accuracy of the election results, and then over plans to return expropriated property.

Nicaragua's other major problem is an upsurge in violence. Gangs of gunmen, known as the *rearmados*, who are former Contras or former Sandinista soldiers, have been terrorising parts of the countryside. President Alemán has agreed to negotiate with them, offering land, equipment, financial assistance and building materials. This is perhaps the last chance to negotiate an end to a small, but resurgent armed conflict that began as a political struggle but has degenerated into little more than armed banditry.

Panama continues to debate the future of the Panama Canal. President Ernesto Pérez Balladares insists that the Canal will be returned to Panama in 1999 as stipulated in the Panama Canal Treaty and the Treaty on the Permanent Neutrality and Operation of the Canal, both signed by US President Carter and Panama's President Torrijos on 7 September 1977. However, the US has proposed converting one of its military bases, Howard Air Force Base, into a multinational Centre for Drug Control. This base is already used as a centre for anti-narcotics and surveillance missions, including hundreds of US military flights over the Andean, Amazonian and Caribbean regions. Many Latin American nations have expressed their opposition to this plan, while many Panamanians see it as a way to maintain a US military presence in their country, even though it violates the

Canal treaties. As 1999 approaches, this issue is likely to intensify, not only in Panama, but throughout the region because of the symbolic and charged nature of both the Panama Canal in Latin American history and the current political divide over the war on drugs.

Concomitantly, Panama is beginning to feel pressure along its southern border, as Colombian guerrilla violence and the coca/cocaine trade spill across the frontier in the Darién Straits. The Colombians, for their part, have expressed concern over sightings of US soldiers in the Darién region. The Panamanian government has proposed creating a special defence force to meet the new security needs. If such a force were created, it would reverse the de-militarisation trend that followed the 1989 US invasion, when Panama's armed forces were dismantled and a police-style Public Force created.

Panama occupies an ambiguous position within Latin America's current security environment. It is now viewed as a drug-producing, trans-shipment and money-laundering haven, yet it is also the regional beachhead for the US war on drugs. If Howard Air Force Base becomes a drug-control centre, Panama will remain in the forefront of hemispheric security issues, even as the US Southern Command relocates to Miami.

The Caribbean, too, is increasingly viewed by US security agencies as harbouring the drug trade. From Jamaica to Haiti to Cuba, the US is concerned by the impact of large, illegal trading practices on small and vulnerable nation-states. Most of the small nations are cooperating with the US: even Cuba has expressed its willingness to work with the US, and the Cuban Coast Guard has on occasion cooperated with its US counterpart.

Cuba, however, is not seen by the US as a potential ally in the drug war; it receives no US aid and thus is not subject to certification. Instead, Washington continues to view Cuba as a Cold War-style security threat despite its profound economic crisis. Even though Cuba is opening up to foreign investment and markets in an effort to stabilise its economy, the US steadfastly refuses to rethink its Cuba policy. This continued rigidity has begun to create conflicts between the US and its Canadian, European and Mexican allies who have taken their case against the US Congress' 1996 Cuban Liberty and Democratic Solidarity Act (commonly known as the Helms–Burton Act) to the WTO. Although most of these countries have stated that they favour a democratic transition in Cuba, they believe that US policy is counter-productive. Once again, Cuban President Fidel Castro has managed to win much of the world's support for his country's long-term opposition to US sanctions.

Troubles in the Mountains

Drug-related violence, US narcotics policy and impressive rebel activity continue to challenge the Andean-region governments of Bolivia, Colombia, Ecuador, Peru and Venezuela. Colombian guerrillas mounted major offensives during 1996, demonstrating that their military capability is greater today than at any previous point in the last 30 years. In Peru, *Sendero Luminoso*, or Shining Path, has stepped up its military actions during the past year, despite its weakened condition since the capture of its leader, Abimael Guzmán, in 1992. The nation's other guerrilla movement, the *Movimiento Revolucionario Túpac Amaru* (MRTA), long overshadowed by *Sendero*, staged a spectacular takeover of the Japanese Ambassador's residence in Lima on 19 December, initially holding 700 people hostage including more than 20 ambassadors and top-ranking Peruvian government and military officials.

Even without guerrilla actions, democratic institutions in the region have come under great strain. Peru's party system has not survived the blow dealt by President Alberto Fujimori's 'self-coup' when he shut down Congress in 1992. Despite the elections – and Fujimori's re-election – in 1995, Peru's traditional parties have all but disappeared, receiving less than 5% of the vote in recent elections. Fujimori seems disinclined to convert his campaign vehicle, Change 90, into a political party. There thus appears to be an institutional vacuum, with Fujimori on the top, the population on the bottom, and few institutional mechanisms of governance in between.

The situation in Ecuador appears equally unstable. In February, Congress assumed the authority to overthrow President Abdalá Bucaram Ortiz, citing a clause in the Constitution concerning his 'mental incompetence'. This action threw Ecuador into an institutional free-for-all, with three claimants vying for the presidency. The President of Congress, Fabian Alarcón, was ultimately selected for a two-year period. Throughout, the presence of the armed forces, whose support was decisive in settling the institutional crisis, was palpable and public.

Yet even as traditional democratic institutions faltered in some areas, innovative ones advanced in others. Most notably, long-marginalised indigenous communities continued to increase their participation and access to state resources at the national level, gain greater self-governing authority at the local level and increased legal protections for cultural pluralism and ethnic diversity in general. This is particularly true in Bolivia and Ecuador, nations with approximately 70% and 40% indigenous populations respectively. In Colombia, the push for greater multi-ethnic, multi-cultural rights has also led to greater enfranchisement of its large African-Colombian population, concentrated primarily on the Atlantic and Pacific coasts and in the mountainous area south of Cali. This greater incorporation of diverse populations in the Andean region – as well as in Guatemala through the peace accords – represents a silent revolution in Latin America, leading to sweeping changes in the way the nation-state relates to culturally diverse civil societies.

These competing and contradictory trends have had differing effects in the five nations of the Andean Community. There is great tension in

Colombia between the subversive effect of drug-traffickers and their associates, and US pressure for a more concerted assault on cartel leaders and the illicit drug economy. Following the precipitous decline in US–Colombian relations in the wake of allegations that President Ernesto Samper's 1994 electoral campaign relied heavily on drug money, the situation worsened when the US de-certified Colombia in March 1996. Shortly thereafter, President Samper's US visa was revoked; the only other US visa revocation of a democratically elected leader was that of former Austrian President Kurt Waldheim as a result of his earlier links to Nazi war crimes in the Balkans.

Since de-certification, the US has increased its anti-narcotics assistance to the Colombian military, despite continued documentation of major human- rights violations. Moreover, with US backing, the Colombian armed forces began a major campaign to eradicate coca fields, directly confronting sizeable guerrilla armies in these areas and pitting troops against hundreds of thousands of farmers who have few economic alternatives to coca farming. The response was predictable: after chemical spraying and forced eradication, tens of thousands of farmers joined in massive demonstrations, blocking roads and marching on regional capitals. The government sent negotiators and agreed to work more closely with the communities and halt the chemical spraying. Yet once the protests subsided, the military continued its offensive, revealing a weakened President with diminished authority over his armed forces. This may be the greatest crisis in Colombia.

In March 1997, despite major crop-eradication efforts, the prosecution of drug cartel leaders and the revision of laws to expropriate illicitly gained property, the US again de-certified Colombia, citing official distrust of President Samper as the justification.

In Peru, the US has developed a good working relationship with President Fujimori. Despite allegations and evidence of high-level, drug-related corruption and the systematic suborning of mid-level military officers in the drug-producing zones, Peru was easily certified in 1996. Peru (together with Colombia) has cooperated in installing radar stations that have effectively shut down most of the illegal air traffic linking the coca fields of Peru and Bolivia with cocaine laboratories in Colombia. During the past year, about 30 planes were shot down or intercepted by the Peruvian Air Force with the collaboration of US Southern Command. When the traffickers responded by opening land and river routes, the US authorised a major investment in riverine patrols in Peru as a counter-measure. As a result, more of the traffic now flows across the borders of states such as Brazil, Ecuador and Paraguay.

The Peruvian military has been more successful than its Colombian counterpart in reducing the size and threat of guerrilla insurgency. Yet it may have become too complacent. Over the last year, a dissident faction of *Sendero* has increased attacks, particularly in the Upper Huallaga and

Apurimac Valleys, centres of the coca trade. The faction, known as *Sendero Rojo* (or Red Path) has ignored the call by imprisoned guerrilla leader Guzmán to suspend military actions and take up politics.

On 19 December 1996, the MRTA seized the Japanese Ambassador's residence during a reception honouring Emperor Akihito's birthday. The guerrillas took hostage 20 ambassadors from Europe, Asia, North and South America, together with Peruvian Foreign Minister Francisco Tudela, President Fujimori's brother and senior military and civilian leaders. During the next three weeks, the MRTA released most of the captives, reducing the number of hostages to about 70 Peruvian officials, Japanese diplomats and businessmen and the Bolivian Ambassador. The guerrillas demanded the release of over 400 of their political prisoners from Peruvian jails. President Fujimori refused; however, under Japanese pressure, the hardline President agreed to seek a negotiated settlement. In February, following a visit by Fujimori to Havana, President Castro agreed to grant asylum to the guerrillas if a negotiated settlement was reached. Yet, on 15 March 1997, three months after the initial takeover, the situation remained unchanged.

In Ecuador, populist Guayaquil politician Abdalá Bucaram Ortiz upset conservative Jaime Nebot Saadi in run-off presidential elections on 7 July 1996. In the earlier parliamentary elections, Bucaram's party, the *Partido Roldosista Ecuatoriano*, won only a minority of the seats while a coalition of Ecuador's indigenous population movements running in national elections for the first time won seven Congressional seats, a major breakthrough.

Bucaram's election raised uncertainty about Ecuador's future and heightened expectations that the country would not join the regional trend towards neo-liberal reforms. During his campaign, candidate Bucaram called for 'a better distribution of wealth and equitable development' and proposed spending more than $2bn on social programmes. Yet once in office, President Bucaram instituted a harsh, economic adjustment programme which in turn led to major street protests by labour and indigenous movements. By January 1997, a broad spectrum of the population joined the protest against the economic measures. In February, the Catholic Church, a group of ex-Presidents and other prominent politicians voiced support for a national strike. Only five months after Bucaram had won a presidential run-off by 54% and taken office, a national public opinion poll revealed that he had the support of less than 25% of the population.

The growing unrest sparked Congressional action; the unicameral legislature declared the President 'mentally incompetent' and stripped him of power on 5 February. On 11 February, Congress elected its own President, Fabian Alarcón, as interim President. The move was contested by both Bucaram and his Vice-President, Rosalia Arteaga. The Constitution was ambiguous about the right of succession. The armed forces had initially backed Arteaga – who was sworn in on 9 February, but only held office for two days – and then supported Alarcón once Congress agreed to clarify the succession issue. The incident revealed the lack of social or institutional consensus on such fundamental issues as the role and legitimacy of democratic institutions, and the direction of economic policy. Ecuador, whose democratic transition in 1979 was the first in the region, has still not been able to consolidate a stable, institutional framework for democratic governance.

Bolivia, like the other Andean nations, has been caught in the crossfire between drug trafficking and US anti-narcotics policies. With US help, special military and police units have been created to eradicate coca in the Chapare, the tropical zone east of the Andean city of Cochabamba. In March 1996, the military established a jungle training school at Chipiriri to be supervised by a specially trained police unit, the Special Force for the Fight Against Drugs Trafficking (FELCN). Coca growers denounced the move as a militarisation of the region. Nevertheless, Bolivia announced that all coca will be eradicated from Chapare by 2000. As coca eradication has increased – and Bolivia easily reached US goals – national and international human-rights groups have denounced an increase in human-rights violations.

Despite the contentious nature of the drug trade in Bolivia, the government of Gonzalo Sánchez de Lozada, universally known as Goni, has implemented some major reforms to advance the democratisation process and completely restructure the economy. Goni's administration passed a law allowing for the direct election of mayors in 311 newly created municipalities. The law also gives 20% of state revenues to the municipalities, distributed according to population. Even the coca farmers have formed their own political parties and elected their candidates as mayors of the three principal municipalities in Chapare. Also in place is the Agrarian Reform Law, declared on 17 October 1996, which for the first time recognises indigenous territory and establishes a separate class of rights, responsibilities and cultural protections. The new laws have led to the election of indigenous representatives at the local level, extending the electoral breakthrough of Vice-President Victor Hugo Cárdenas, an Aymará Indian.

Economically, the Sánchez de Lozada government has opened up the state sector to private investment, selling 50% of the stock while giving 100% control to the private sector and foreign investors. The move has led to major investments in the energy and mining sectors, potentially transforming Bolivia into an energy hub for southern Brazil and the southern cone countries of Argentina, Brazil, Chile, Paraguay and Uruguay. Bolivia also negotiated an 'association' agreement with Mercosur (*Mercado Commun del Sur* – the Southern Cone Common Market) to provide early access to the Mercosur market ahead of its Andean Community (the

Andean trading area) partners, who are collectively negotiating with Mercosur. The Bolivian agreement took effect on 1 January 1997. Much of the dramatic economic, political and social transformations taking place in Bolivia can be traced to an unusual alliance between indigenous groups, represented by Vice-President Cárdenas, and neo-liberal technocrats, represented by Goni.

Venezuela's transformations have been less smooth. It has the largest debt in the Andean region and its economy remains mired in economic recession. On 16 April 1996, President Rafael Caldera Rodríguez announced an economic shock treatment, known as 'Agenda Venezuela', which liberated the exchange market, enacted price controls on wages, removed transportation subsidies and increased sales taxes. The measures were part of an agreement with the International Monetary Fund (IMF). The agenda also includes a major programme to privatise state enterprises.

Inflation remains inordinately high at 103% annually in 1996. With the shock treatment, government ministers predict the rate will fall to below 50% in 1997. These rates still diverge greatly from Venezuela's neighbours which have mostly brought inflation under control. The programme will also require a massive restructuring of the country's public administration. The plan calls for reducing the number of state employees from 1.3m to 800,000. The austerity measures have triggered innumerable strikes, particularly by public-sector employees whose jobs are directly threatened.

On the security front, Venezuela is not directly in the crossfire of the drug war. However there is much talk and fear of the 'Colombianisation' of Venezuelan society. Police brutality and extra-judicial killings have risen. In Caracas, the murder rate has increased to 60 per 100,000, one of the highest in the world. There have been repeated incidents of Colombian guerrillas crossing the border into Venezuelan territory and attacking police and military posts. Three Colombian guerrillas and seven collaborators captured in Venezuelan territory in July 1996 were tried in military courts. The violence and instability along the Colombian–Venezuelan border continues to create tension between the two countries.

The Southern Cone and Brazil

The southern cone has not been a central focus of the US war on drugs. Yet its ports, airfields, territory and financial institutions are increasingly used for trans-shipment and money laundering. In Chile, 600kg of cocaine were seized in the San Antonio port, leading to a major revision of its anti-narcotics policy. Chile has begun to increase border controls and to construct physical barriers along key international routes. Uruguay, too, has become a major drug trans-shipment and financial centre. Montevideo has emerged as a money-laundering hub due to its strict banking secrecy laws. Yet, despite a growing presence in the region, the illegal narcotics trade has not had the destabilising effect on democratic institutions and civil–military relations that it has had in the Andean region. The principal threat to democratic stability in this region still comes from the legacy of military power that characterised the democratic transitions in the 1980s.

The most striking incidence of military insubordination occurred in Paraguay, when on 22 April 1996, President Juan Carlos Wasmosy asked for the resignation of senior Army commander General Lino César Oviedo Silva. The General refused, triggering an international crisis that threatened to disrupt a hemispheric trend towards consolidating civilian democratic rule. Since 1991, the Organisation of American States (OAS) has created new multilateral instruments to enforce its commitment to representative democracy in the hemisphere. Chief among these is the 'Santiago Resolution' which requires the convocation of the Permanent Council of the OAS when democratic government is interrupted in any member-state. Prior to the Paraguayan case, the Santiago Resolution was invoked during the 1991 coup in Haiti, the 'self-coup' in Peru and a similar attempt in Guatemala in 1993.

Forceful action by the US Department of State and the OAS, together with a massive show of support for Wasmosy through street demonstrations, turned back the coup and forced the General's resignation. The intervention of other Mercosur nations also signalled a new diplomatic actor and regional framework for stable, democratic rule in the southern cone and Brazil.

In Chile, civilian and military authorities continue to clash over constitutional prerogatives and institutional roles. Proposed reforms would have eliminated the appointed, lifetime senators and given greater congressional and presidential control over military retirements. They also would have reduced the power of the Pinochet-dominated National Security Council. The balance of power in the Senate, however, where 15 Senators (some appointed for life) are loyal to General Augusto Pinochet, makes constitutional reform impossible.

The civilian government of President Eduardo Frei Ruiz-Tagle has also been frustrated by the failure to punish human-rights violations from the Pinochet period. The escape of four leaders of the *Frente Patriótico Manuel Rodríguez*, a guerrilla organisation linked to the Chilean Communist Party that was active in the 1980s, also underscored the breakdown in civil–military relations. The political prisoners were freed in a dramatic helicopter rescue backed by heavy artillery. Civilian and military authorities blamed each other, yet President Frei's attempt to extend his power and reorganise the state security apparatus was rejected by Congress.

The civil–military conflict has resulted in Chile moving against regional trends to reduce military expenditures. While Argentina has steadily cut its military's size and budget, Chile's military payroll has doubled since the 1970s. Its per capita defence spending is now twice that of Argentina. Argentina has tried to re-orient its armed forces towards participation in international peacekeeping missions to match Chile's long-standing involvement in such operations. Moreover, the military continues to be directly involved in the economy, taking a percentage of revenues from the sale of copper and controlling defence-related industries.

Despite these difficulties, the Chilean economy, which grew 6.9% in 1996, continues to be the most stable and buoyant in Latin America. Chile is also a candidate to become the first South American entrant into the North American Free Trade Association (NAFTA), and has independently associated itself with Mercosur.

In Brazil, drug trafficking is a more central issue. This nation, which covers over 58% of the Amazon Basin and is the largest South American country, has been converted into a producing, trans-shipping and major consuming nation in the global drug trade. Brazil has resisted being drawn into US-dictated anti-narcotics policies and thus far has avoided many of the conflicts with the US that have unsettled the Andean nations. During 1996, Brazil turned down $1m in anti-narcotics assistance because it disagreed with the US approach. President Fernando Henrique Cardoso also refused to involve the Brazilian military in a joint anti-drug task force. Yet Brazil's low profile within the world narcotics trade is likely to change, and that change will bring closer scrutiny and more conflicts with the United States.

Historically, the Brazilian military's principal mission was to be prepared for Argentine aggression and to fight internal communist subversion. Today, Argentina is a close ally and trading partner and the armed forces of the two countries engage in joint exercises. The end of the Cold War has minimised the emphasis on internal subversion.

In October 1996, President Cardoso announced a new national defence policy explicitly giving the President, not generals or admirals, control over all aspects of military power from strategic planning to implementation. The centrepiece of Brazil's new security policies focuses on defending the vast empty spaces of Amazonia. New military bases will be built throughout the region and the military will be responsible for extending social programmes, such as health and education, into the interior. This action is designed to resolve the conflicts caused by social inequality, poverty and ethnic divisions, and to halt the spread of international drug and arms trafficking in this vast territory, which is emerging as a strategic site for illicit activities as a result of crackdowns in the Andean region.

To implement this security policy, the Brazilian Congress approved the use of the *Sistema de Vigilencia de Amazonia* (SIVAM), a monitoring system using land- and space-based radars. Despite major concerns about external involvement in Brazilian national security, the Senate approved the project which will be financed by the US Export–Import Bank and a consortium headed by the Raytheon Corporation.

Brazil's economy has stabilised under President Cardoso, although it still suffers from a large fiscal deficit and has signed an agreement with the IMF to reduce public spending by $6.5bn and to reduce the deficit from 3.5% of gross domestic product (GDP) to 2.5%. Yet the government seems reluctant to institute full austerity measures, largely because of the nation's great social problems. According to a UN report, Brazil has the greatest income inequality in the world. Brazil's richest 10% of population holds 65% of the national wealth, while the poorest 40% have only 12%.

There is still great pressure to distribute land to the rural poor, particularly in the north-east and in parts of the Amazon Basin. Rural conflict has risen since 1995. Cardoso promised major land reform at the outset of his term, but he has not delivered. Instead, land invasions have increased in size and number, as have related political and social violence. With mounting pressure from the right, Cardoso recently threatened to treat future land invasions as a national-security issue and to deploy the army to evict transgressors. In September 1996, a law was introduced in Congress to hasten the distribution of lands from 35,083 unproductive estates with more than 1,000 hectares each. The unproductive estates cover a total area larger than the combined territories of Austria, France, Germany, Spain and Switzerland.

The government has also come into conflict with the indigenous communities of the Amazonian Basin. The President issued a decree which undermines an earlier law passed by former President Fernando Collor providing for the demarcation and legal recognition of indigenous lands. The new decree gives landowners and farmers legal recourse to defend their interests. The indigenous community claims this measure could drastically reduce their territories and has vowed to resist further encroachment. The issue went to the Brazilian Supreme Court which upheld Cardoso's decree. With international non-governmental organisations (NGOs) pledged to support the indigenous populations' claims, these rights are being transformed from social to national-security issues, and will continue to place a severe strain on civilian control of the armed forces, and lead to more social unrest.

Mexico: The Troubled Land

Mexico continues to confound optimists who predict its smooth integration into the North American economy and a slow, but steady advance towards pluralism and democratisation. Only two years ago, President Carlos Salinas de Gortari was feted as a hero of both economic modernisation and political liberalisation. Today, he is a discredited man living in exile in Ireland. His brother, Raul Salinas, is in jail, charged with corruption and possible involvement in drug-trafficking and high-level political assassinations. Shortly after Salinas left office in August 1994, Mexico's peso was sharply devalued, triggering a major economic crisis from which the country has still not recovered. The expected gains of NAFTA have not been translated into real benefits for the majority of the people, and the economy in 1996 remained mired in recession. Although exports increased by 50% since 1993, the year before NAFTA came into effect, per capita growth was 4% lower in 1996 than in 1993.

President Ernesto Zedillo appears committed to political and economic reform; he is continuing the trend of allowing more political space to the two opposition parties, the conservative *Partido Acción Nacional* (PAN) and the centre-left *Partido de la Revolución Democrática* (PRD). Reforms were introduced in Congress that would increase the opposition parties' access to television, enforce limits on campaign expenditures and reduce opportunities for electoral fraud. Yet congressmen from President Zedillo's *Partido Revolucionario Institucional* (PRI) failed to support the reforms. Nevertheless, many expect that the ruling PRI will lose its majority in Congress for the first time during the next Congressional elections in July 1997. In 1996, the PRI lost control of congresses in the states of México and Coahuila, as well as local offices in Michoacan, Chiapas, Puebla, Sinaloa, Oaxaca and Coahuila. The PAN is now expected to win the mayoral race in Mexico City, and the PRD expects to win the governorship in Campeche. Quite evidently, the PRI's monopoly on power is breaking down, auguring well for democracy.

Yet, as the failure of the political reforms suggests, much of the party apparatus seems unwilling to stand by quietly as 68 years of one-party rule is threatened. The PRI is divided between the young technocrats, like President Zedillo, willing to promote economic and political liberalisation, and the 'dinosaurs', traditional party bosses accustomed to a one-party authoritarian regime. Many now accuse the dinosaurs not only of obstructing reforms, but also of using violence to preserve their power. Elements of the PRI, perhaps with the complicity of Raul Salinas, have been connected with the murders of PRI presidential candidate Luis Donaldo Colosio in 1994 and other high-level assassinations. The PRI is today a divided party; it is a very uncertain overseer of political and economic modernisation.

While the question of whether Mexico can democratise peacefully is prominent, security issues are increasingly on the agenda, as well. One of the keys to the PRI's longevity was the firm civilian control it maintained over the military. This is now fraying. Three years of peace talks between the Zapatista rebels in Chiapas and the government have led to little progress, despite the guerrillas' declaration that they would like to transform from an armed, insurgent movement to a legal, political movement. In the absence of peace, President Zedillo has placed military officers in charge of internal security and in key police roles. The military's internal security role has intensified with the rise of a second guerrilla movement, the *Ejército Popular Revolucionario* (EPR). In August 1996, the EPR undertook a coordinated assault on police and military posts in the states of Guerrero and Oaxaca (see map, p. 236).

Yet the principal factor contributing to the breakdown of internal order and civilian control over the military is the decision to deploy the armed forces in the war on drugs. Before Zedillo took office, Mexico's military had mostly played a supporting role in the nation's anti-narcotics policies. They were responsible for eradicating marijuana and opium poppy fields. Yet with evidence of widespread corruption in the federal and state police – and with the major backing of the US – President Zedillo began to place military men in key positions to fight the drug cartels. General Jesús Gutiérrez Rebollo was appointed Mexico's chief anti-drug official in December 1996. Similarly, Army officers were placed in charge of municipal police forces in principal transit states, such as Sinaloa, and in the border state of Baja California. Just as in the operations against Mexico's guerrillas, Zedillo has broken with the past and begun to politicise the long-subordinate military.

The results have been disastrous. Zedillo's decision violates a cardinal lesson in drug-war politics, learned from experience in the Andean region: deploying armed forces in the drug war inevitably corrupts them. This was brought home when General Gutiérrez was arrested for collaboration with Mexican drug traffickers on 18 February 1997, only days before US certification was announced. Gutiérrez was privy to a wealth of intelligence about US anti-narcotics operations in Mexico. He had also just returned from a successful trip to Washington where he was widely praised for his anti-drug efforts. He was subsequently linked to a number of drug-related kidnappings.

The arrest of Gutiérrez threatened to derail Mexican certification, as US Congressional members openly called for Mexico's de-certification. Yet the incident revealed how differently the US handles its foreign relations with Mexico and Colombia. For Colombia, virtually all the bilateral relationship is reduced to the issue of drug trafficking. In contrast, US relations with Mexico are multi-faceted and cover a range of vital interests, from trade and immigration to environmental issues along a shared border. The US was unwilling to jeopardise relations with its third largest trading partner over drug-trafficking. The Clinton administration argued that it was important not to undermine a stable working relationship between the two nations. The arguments were convincing, but could just as easily have been applied to Colombia.

Nevertheless, Mexico has been placed on warning: with an estimated 70% of the cocaine headed for the US passing through Mexico, Washington will no longer overlook the drug issue in its bilateral relations. These are headed for future conflict at a time when Mexican internal politics are increasingly unstable. A growing drug presence may hasten the decline of the PRI and lead to a period of general instability within Mexican politics. It is far from certain that in these circumstances the PAN and the PRD would be able to fill the political vacuum and restore order and legitimacy. Ultimately, the issue of democratisation in Mexico is linked to national security.

The Vital US Link

Most nations in Latin America continue to work towards free trade in the region and for greater access to the US market. The cause of regional integration received a boost during a visit to Washington by Chilean President Frei in February 1997. Then, the Clinton administration announced that Chilean admission to NAFTA would be given high priority. Earlier in the year, Mercosur and the Andean Community began talks to accelerate the creation of a South American free-trade area. Past negotiations had been principally on a bilateral basis.

Yet security issues continue to intrude in this area, too. One source of tension was the passage by the US of the Helms–Burton Act – penalising foreign companies for investing in Cuba – in March 1996, following the shooting-down of two unarmed civilian aircraft piloted by Cuban-Americans. The law penalises foreign companies for investing in Cuba. US allies, including Mexico and Canada, claim that the law is a violation of both the General Agreement on Tariffs and Trade (GATT) and NAFTA and that it is an extra-territorial application of US law. The case was taken to the WTO which named an arbitration panel in January 1997. The US immediately declared that this was a security not a trade issue and would not recognise the jurisdiction of the WTO over US national-security policy.

A coherent framework for inter-American relations in the post-Cold War environment is still not well developed. US–Cuba relations remain a lonely vestige of the historic conflict that once shaped global politics. The issues of economic integration, drug trafficking and democracy have rapidly moved to the top of the new hemispheric agenda. During 1996, the opportunities, risks and pitfalls of the emerging agenda were repeatedly underscored.

Europe

Western Europe had set itself a daunting schedule of tasks for 1996 and 1997: to complete preparations for Economic and Monetary Union (EMU); to advance the Inter-governmental Conference (IGC); and to enlarge both the European Union (EU) and NATO. Not surprisingly, it proved difficult to make significant progress on any of these ambitious projects except for NATO enlargement. With the US taking the lead, Europe was on target to invite a number of former Warsaw Pact countries to negotiate entry into NATO in 1997.

Predictably this provoked an outcry from Russia, where a consensus across the political spectrum saw enlargement of the Alliance as a threat to Russia's security. This was the only point of agreement within the country, however. Russia has yet to deal with its profound political and economic challenges following the debilitating illness of its re-elected President Boris Yeltsin, the consequent lack of strong direction, and increased crime, corruption and the fragility of the still-developing democratic system. Yet, by March 1997 there were indications that improvement was on the way. Yeltsin returned to Moscow looking physically fit and mentally positive, reorganising his cabinet and bringing in young liberal reformers. And the economy, for all its troubles, looked surprisingly promising.

The best news from Europe was that peace in Bosnia was still intact. NATO forces were still enforcing the General Framework Agreement for Peace in Bosnia and Herzegovina (the Dayton Accords), but their size had been reduced with no return to war. While there has been little diminution of nationalist fervour and parts of the peace agreement remain unfulfilled, there was some advance in 1996. Much-needed economic reconstruction is under way and there has been some political effort to address common problems. It is a mixed picture: Bosnia is still a fragile state, but one in which there is more potential than at any time since the break-up of the former Yugoslavia.

A Complex Year in Western Europe

Europe's strategic agenda in 1996 and early 1997 was dominated by four major challenges: Economic and Monetary Union; the Inter-governmental Conference; and NATO and EU enlargement to the East. While Europe's leaders continued to pledge that these projects would all proceed on schedule, doubts remained, exacerbated by continued slow economic growth. With political leaders more concerned about short-term or domestic political necessities, public-relations campaigns to promote the new currency – the euro – and prepare for a series of ratification debates and referendums on the future of the 1992 Maastricht Treaty on European Union and on NATO and EU enlargement were slow to emerge in 1996.

These four major projects remained hostage, moreover, to unemployment averaging almost 11% across Europe. High unemployment exacerbated budget deficits, making the Maastricht Treaty's criteria for EMU more difficult to meet. With average EU economic growth for 1996 at a mere 1.4% – mostly export-driven – reducing unemployment proved extremely difficult. While structural problems lay at the heart of the high unemployment, efforts to achieve the Maastricht criteria prevented public spending from stimulating temporary demand. Indeed, most countries were cutting public spending deeply, although a number of dubious plans were also proposed for one-time transfers to stabilise the 1997 numbers. During 1996, Europe's financial and political élites generally concluded that monetary union would be achieved in 1999. In early 1997, however, a sharp rise in German unemployment to its highest level since 1933, and questions about Chancellor Helmut Kohl's staying power, reawakened doubts. Unemployment consumed political capital and attention that might otherwise have been used to address the challenge of widening and deepening Europe's institutions.

Tracking the Major Players

Nowhere was the connection between economics and foreign policy clearer than in Germany. On EMU, the Kohl government faced the dilemma of having to insist on strict adherence to the Maastricht convergence criteria in order to gain public acceptance for a euro as 'strong as the deutschmark', while running the real risk of not reducing its own 1997 budget deficit to below the 3% mark. With much arm-twisting, Finance Minister Theo Waigel gained EU approval at the 20 December 1996 Dublin European Council for a watered-down version of his 'Stability Pact' to ensure continued compliance (via fines) with the Maastricht criteria on public spending after 1999. The problem had been to arrive at a clear economic definition of 'severe' recession, which it had already been decided would be sufficient reason to run a deficit over the 3% level without penalty. Agreement was finally reached to allow states whose gross domestic product (GDP) dropped by 0.75–2.0% to seek permission from the EU's Council of Ministers to run above the 3% deficit level without penalty. Reigning in public spending proved to be a greater challenge. In April 1996, Chancellor Kohl presented an austerity plan aimed at cutting DM70 billion from the German federal budget. The plan, which drew vigorous opposition, particularly from Germany's labour unions, was finally passed in September. Despite these cuts, rising unemployment continued to increase public spending, making Germany's goal of reducing its 1997 deficit to under 3% highly unlikely.

Deep spending cuts and high unemployment damaged Kohl's popularity in 1996. Nevertheless, there was increased discussion of the Chancellor running for an unprecedented fifth term in 1998, which he himself did not rule out. Kohl's most likely challenger, Gerhard Schröder, leader of the Social Democratic Party (SPD), heavily criticised EMU, raising the possibility that the 1998 election would also be a referendum on the euro, which, according to opinion polls, is supported by less then 40% of the population.

Germany's foreign policy focused on moving European integration forward, often in the form of joint initiatives with France to underline support for EMU and to outline the Maastricht Treaty revisions that the IGC should pursue. But Germany was also active in the countries to its east. In December 1996, following an 18-month deadlock in negotiations over forced expulsions after the Second World War, Prague and Bonn signed a declaration of reconciliation, thus completing a series of agreements that Germany had signed with its neighbours to the east since the end of the Warsaw Pact in 1991. The special nature of the German–Russian relationship returned to the fore in 1996 and early 1997. In September 1996, Kohl became the first Western statesman to visit Russian President Boris Yeltsin after his re-election, and in January 1997 he was the first Western statesman to visit Yeltsin after his heart surgery. Germany had a more difficult time with China. Relations deteriorated after the German parliament in June passed a resolution condemning China's oppression of Tibet. Germany also took the final step towards 'normalising' *Bundeswehr* out-of-area operations when, on 13 December, it approved the deployment of 3,000 German combat troops in the NATO-led Stabilisation Force (SFOR) in Bosnia (the third largest contingent after the US and UK), with most serving in a Franco-German unit deployed outside Sarajevo. What was most remarkable about the *Bundestag*'s decision was how uncontroversial it was; the resolution to deploy troops was passed by a majority of 499 to 93 and was even supported by the SPD and most of the Green Party.

In France, President Jacques Chirac showed that Gaullism and Europe do not have to be mutually exclusive. Chirac, in true Gaullist style, was conspicuous on the world stage, anxious to demonstrate France's enduring roles in the Middle East, Africa and Asia. But he also maintained the pro-European position of his non-Gaullist predecessors, joining with Germany to ensure that the IGC and EMU stayed on course, and bringing France closer to a 'new' NATO. Recognising that a European security identity could not be created outside the Alliance, France participated fully in NATO's June ministerial meetings and supported a landmark agreement allowing Europe to undertake its own military missions with NATO assets. But not all was well on the Atlantic front. Chirac's ostensible affinity for the United States did not prevent a series of disagreements with Washington over the Middle East and Africa, military action against Iraq, the reappointment of UN Secretary-General Boutros Boutros-Ghali, and NATO's Southern Command.

Nor were relations always smooth with Germany. Sharp differences arose in particular over Germany's 'Stability Pact', reflecting an underlying division between the two countries about the degree of political control over the new European Central Bank – particularly in view of over 12% unemployment in France. At the same time, France's continuing budget-cutting efforts, spearheaded by Prime Minister Alain Juppé, increasingly pointed to an annual deficit of under 3% for 1997. Nevertheless, major strikes took place throughout the year, culminating in a 12-day truck-drivers' strike in December, underlining how precarious progress on budget-cutting was.

The UK's traditionally strained relations with Europe were complicated in 1996 by the EU Commission's decision in March to institute a world-wide ban on the export of British beef and beef by-products, and by Prime Minister John Major's increasingly tenuous hold on power. In May, Major retaliated against the Union veterinary committee's extension of the ban. He declared in the House of Commons that the UK would pursue a policy of 'non-cooperation' with the Union until the ban was revoked. Mad cows thus derailed the EU's June summit in Florence, which was supposed to concentrate on the IGC. Nor were the UK's own aims for the IGC particularly cooperative. By espousing a minimalist agenda, Major opposed more qualified majority voting, greater powers for the Commission and the European Parliament, and the merging of the Western European Union (WEU) with the EU.

On EMU, Major's cabinet was split between those who wanted to rule out entry into monetary union during the next parliament and those who wanted to leave the door open. Kenneth Clarke, Chancellor of the Exchequer, was the most prominent advocate of Britain's entry into EMU. The split in the Tory cabinet took on less importance as 1997 unfolded and it became increasingly clear that in all likelihood a Labour cabinet would decide on monetary union. How fast Labour leader Tony Blair would be able to move on Europe, should he win the 1997 general election scheduled for 1 May, was, however, open to question. Whatever its views, a new Labour government would still have to set up an Independent Central Bank before any move towards EMU could be considered. At any rate, both the Conservative and Labour Parties pledged to hold a referendum on EMU before reaching a final decision.

For the first time in Italy's post-war history, a centre-left coalition came to power in elections in April 1996. The Olive Tree coalition, under Prime Minister Romano Prodi, pursued a path of fiscal conservatism, affirming its intention to undertake the difficult budget-cutting necessary to join

EMU in 1999. Although Italy's 1997 budget, finally passed in December 1996, contained significant cuts, many of its neighbours to the north feared that early Italian participation would dangerously weaken the euro. Spain, too, under its new government led by José Maria Aznar, made every effort to fulfil the Maastricht criteria and be among the founding members of EMU – causing the same consternation among those EU partners worried about the euro's strength and credibility.

Rethinking Cooperative Ventures

European states, particularly France and Germany, cooperated on a number of foreign- and security-policy initiatives in 1996 and early 1997. Franco-German security cooperation took on a new operational dimension with the deployment of the Franco-German brigade headquarters to command a joint SFOR unit in Bosnia. At the same time, the two countries attempted to coordinate their positions at the IGC, releasing letters at their twice-yearly summits calling for more majority voting, more flexibility through constructive abstention, and the eventual merging of the WEU into the EU.

A confidential joint security-policy memorandum from the December 1996 Nuremberg meeting of the Franco-German Security Council, leaked in January 1997, drew fire from the French National Assembly, particularly after German Defence Minister Volker Rühe interpreted the paper as implying the 'NATO-isation' of France. While the paper largely reaffirmed previous positions, it went further by, for the first time, explicitly calling for a 'dialogue on the role of nuclear deterrence within the context of European defence policy'. This provoked sharp criticism of the French government by the opposition, demonstrating that there were limits to how far into NATO France might be able to return, even under Chirac.

Franco-German collaboration on arms production was troubled by France's fitful consolidation of its defence industry, but major projects continued, including the *Tiger* helicopter and cooperation on the construction of two costly surveillance satellites, *Hélios* II and *Horus*. On the other hand, work on the Future Large Aircraft (FLA) was put on hold as a result of the restructuring of France's aerospace industry and the tight budgets in both countries.

Germany and France maintained their support for the EU policy of 'critical dialogue' with Iran, despite US criticism (and despite a Berlin court ruling in March 1996 that Iranian leaders were behind a 1992 assassination in a Berlin restaurant). France's role in helping to broker an April cease-fire between the Iranian-backed *Hizbollah* and Israel, and Germany's role in mediating the release of Israeli prisoners by *Hizbollah* in July 1996, convinced many in Europe that the dialogue had its value.

Yet Germany and France also had their differences. In particular, Germany was frustrated at the lack of consultation when France carried out a major restructuring of its armed forces and defence industry. Both Chirac's February 1996 announcement that France planned to end conscription and the decision in July to reduce the number of French troops in Germany from 20,000 to 3,000 took Bonn largely by surprise.

The European Union took a number of foreign-policy initiatives in 1996, which served mostly to demonstrate that it was better at fostering conference diplomacy than at crisis management. In early May, Denmark, Estonia, Finland, Germany, Iceland, Latvia, Lithuania, Norway, Poland, Russia, Sweden and the European Union gathered for a Baltic Seas summit on Gotland Island, Sweden, stressing their opposition to organised crime and environmental degradation, while avoiding touchy security issues like NATO expansion.

NATO's Berlin summit in June 1996 agreed that the North Atlantic Council could henceforth approve WEU use of NATO assets, including the Combined Joint Task Force (CJTF) arrangement, thus marking a milestone in Europe's security identity. While this step went some way towards accommodating France's interest in a more European NATO, it did little to plausibly describe under what scenarios the WEU might act while NATO remained on the sidelines. European forces continued to restructure for out-of-area operations in 1996, but shrinking defence budgets hindered progress towards a true European expeditionary capability.

In February 1997, EU foreign ministers met in Singapore with their Asian counterparts from the Association of South East Asian Nations (ASEAN), China, Japan and South Korea, continuing the Asia–Europe Meeting (ASEM) process started the year before in Bangkok. The initiative begun in November 1995 in Barcelona to increase EU–Mediterranean cooperation continued with a second conference in April, focusing on issues of trade and good governance as well as security. An EU emissary and substantial financial aid anchored Europe's presence in the Middle East peace process.

The EU played a major role in financing reconstruction efforts in Bosnia, after donating 48% of the $1.8bn in international funds thus far pledged. (The US has contributed about 15%.) In April, Ricardo Perez Casado replaced Hans Koschnik as the EU's Administrator in Mostar, but resigned in July, to be replaced by Sir Martin Garrod. Violent demonstrations in February 1996 underlined that reconciliation had not yet come to the city despite the EU's best efforts. European states remained wary of a more independent military role in Bosnia, insisting that their troops would stay in Bosnia only as long as US troops. Memories of the UN Protection Force (UNPROFOR) experience, when the US made policy from afar while the Europeans supplied the troops on the ground, counselled against ever repeating this division of labour. Nevertheless, European NATO members provided SFOR with 14,680 ground troops as opposed to 8,500 from the US.

European armaments cooperation suffered from the declining defence budgets and overdue production. Delays continued to disrupt a final production decision for 620 *Eurofighters*, originally intended for Spain (87), the UK (232), Germany (180) and Italy (121). In a disagreement with the United States over the Helms–Burton Act penalising companies for trading with Cuba, and in particular, over the D'Amato Act that imposes mandatory penalties against foreign companies investing in Iran and Libya, the EU took action in October. The Council agreed unanimously on anti-boycott legislation making it illegal for companies to comply with the Act and granting them the right to counter-sue the US in European courts. The EU also prepared a case against the US in the World Trade Organisation (WTO), although the US argues that Article 21 of the General Agreement on Tariffs and Trade (GATT) allows it to withhold trade against another member on national security grounds. The EU and the US did cooperate, however, to support WTO action in December in Singapore intended to open world markets for information technology. The transatlantic market also benefited from progress in 1996 on mutual recognition agreements regarding product standards.

When it came to managing specific foreign-policy crises, Europe demonstrated less assertiveness. This was particularly clear with regard to Greek and Turkish sabre-rattling in the Aegean. The United States had, once again, to intervene to prevent the Greeks and Turks from coming to blows, this time in January over a Cypriot decision to purchase Russian-made surface-to-air missile 300 air-defence systems. Greece, itself, continued to demonstrate that one country could cripple a European Common Foreign and Security Policy (CFSP) by blocking an EU aid package to Turkey in conjunction with the new EU–Turkish customs union. Greece thus stood in the way of a more active European role in an area of significant strategic importance to it, despite pleas by the new US Secretary of State Madeleine Albright for closer EU–Turkey relations.

The European Union's steady progress towards monetary union also had strategic implications. Although efforts were made to ensure the economic fundamentals were in order, thus ensuring public and market support for the euro, EMU's ultimate purpose was more than economic. It was about taming national political power in an increasingly integrated Europe. It was also about the vision that Europe would one day become a strategic actor in its own right, with monetary union providing a decisive impetus towards this goal.

Preparing for the Future

At one level, the vision of a Europe capable of strategic action also guided the European Union's Inter-governmental Conference, which began in March 1996 in Turin, Italy. But the IGC also faced the more concrete challenge of revising the 1992 Maastricht Treaty to prevent enlargement causing institutional gridlock, and of doing so without alienating the European public, already wary of any bid by Brussels to seize power. The institutional consolidation of Europe's new geopolitics remained entangled in competing objectives: how to reform institutions originally designed for a Community of six states to sustain a Union of 20 or more; how to streamline EU decision-making without compromising national sovereignty; how to buttress the single market with a common currency without creating new divisions between those inside and outside the EMU zone; how to enlarge to the east without bankrupting the EU's structural funds and Common Agricultural Policy; and how to bring new states into the Western security community without antagonising Russia. Shaping all these issues was the question: where do the borders of the new Europe lie? Answering this was made more complicated by the fact that it was not just a cultural or economic issue, but also a strategic one.

Underlining the desire to move quickly ahead with the IGC, the European Council held a special session in Dublin in October, yet progress was slow in the face of a number of vexing issues. Indeed, many believed that the slow pace and the UK elections on 1 May would delay completion of the Conference until late autumn 1997. The main aim of the IGC – to streamline institutions already straining under a membership of 15 and sure to seize up under a membership of 20 or more – implied significant changes to the Treaty on European Union. One of the key issues was the relative voting power of the Union's smaller members. Unless some adjustment was made, a handful of small countries representing a fraction of the EU's population would have more votes than the larger countries with a majority of the population. This was true both of qualified majority voting in the Councils and in the European Parliament. Small countries also faced the prospect of losing the right to have a permanent Commissioner. The European Commission was already top heavy with 20 Commissioners; giving each new member at least one Commissioner while preserving the number each old member had would be impossible. The relative roles of the Council, the Commission and the Parliament were also controversial, with Euro-federalists favouring a stronger role for both the Commission and the Parliament, and Euro-sceptics wary of further diluting national sovereignty and thus wanting the Council to remain the locus of EU decision-making power.

Overarching all of these questions was the issue of flexibility. Majority voting, it was clear, would not be extended to many new areas. Germany and France thus introduced another approach. They sought to anchor into the Maastricht Treaty the right of those members willing to proceed on a particular policy to do so on their own with the others 'constructively' abstaining. Importantly, they wanted the institutions of the European Union, including Finance, to be available for such 'enhanced cooperation' among smaller groups. Such an arrangement would weaken the power of veto without compelling countries to carry out EU policies they opposed. Yet even this alternative to more majority voting found opposition from Denmark, Portugal, Sweden and the UK, who were wary of moves to do away with unanimity, fearing the creation of an inner core of countries that could set the rules without the other members.

On Justice and Home Affairs, the Union remained divided between those concerned to bring immigration and asylum issues, as well as police and customs affairs, under the Commission's roof, and those who wanted to keep these matters in a purely inter-governmental framework outside the EU. Expanding the jurisdiction of Europol and the European Court of Justice also remained controversial. Many hoped that their own frustration with Europe's actions in Bosnia and the general lack of unity and influence demonstrated by the CFSP would encourage the 1996 IGC to enhance the EU's foreign- and security-policy role significantly. Yet the revisions to the Maastricht Treaty's CFSP section (Title V, Article J) proposed by the Irish Presidency in its December 1996 draft indicate that majority voting, at any rate, will not be the means to achieve the sought-after effectiveness.

Constructive abstention, however, appears likely to play a larger role in the Union's future CFSP. Unanimity would officially still be required for Joint Action, but the Treaty would now declare that abstention should not hinder the adoption of such action. Too many abstentions (25 votes in the weighted voting system) would, however, prevent its adoption. More qualified majority voting in implementing, as opposed to adopting, Joint Actions remained possible in early 1997, but it was likely that defence and security actions would remain subject to unanimity in their entirety. A continuing problem would be defining more fully the distinction between a Joint Action and its implementation, since differing views could be exploited by those wanting to assume a particular policy.

The member-states also agreed on the need to create some form of CFSP secretariat to support the Presidency with policy planning and early warning. A Secretary-General of the Council would lead this new facility, assisting the Presidency, but also giving more visibility to the EU's foreign-policy role. Some, however, thought this visibility should not go too far, and rejected the idea of an EU 'foreign minister'. The role of the Commission also remained controversial, but revisions will probably be made to associate the Commission more closely with the Council on CFSP.

As its second pillar, the Maastricht Treaty intended the Common Foreign and Security Policy to be inter-governmental and outside the purview of the Commission. But, over the years, the Commission has developed its own 'foreign policy', with four Commissioners devoted to various aspects of external relations. Hoping to enhance its role even more, the Commission proposed a new CFSP troika of the Presidency, the Secretary-General and the Commission. (The current troika consists of the previous, serving and next Presidencies.)

Where the Commission does have clear foreign-policy authority is in the area of commercial policy. Treaty revisions are likely to extend this negotiating authority – depending on Council approval – to include services, intellectual property rights and direct foreign investment in institutions like the WTO. Revisions to section J.4 on the Union's security and defence role in 1996 had not progressed much beyond more forcefully affirming the aim of eventually moving towards a common defence and, in the meantime, of working more effectively with the WEU. While NATO and the WEU seemed to be developing a clearer relationship, particularly following the June 1996 Berlin ministerial meeting, the arrangement between the two Unions remained controversial. There were those, especially in the Benelux countries, France, Germany, Greece and Italy, who argued that in the long term the WEU should be absorbed entirely into the EU. Yet others, both in the UK and the formally neutral Austria, Finland and Sweden, saw a problem with making EU membership synonymous with the kind of hard security guarantee represented in the WEU's Article 5, particularly with a host of new members about to join. In the short term, France and Germany led a move to give the EU greater authority to 'instruct' the WEU to carry out particular actions. Others opposed this, wanting to keep the European Council out of direct involvement with defence policy. By the end of 1996, the Irish, Swedes and Finns appeared to support yet another formulation whereby the EU could have recourse to the WEU.

It nevertheless seemed likely that the IGC would write the WEU's Petersburg tasks (humanitarian and rescue missions, peacekeeping tasks and the use of combat forces in crisis management) into the Treaty as areas falling under the CFSP. A final, and relatively uncontroversial change, would endow the Union with a legal personality similar to the one held by the Commission. This would allow the Union to negotiate agreements on CFSP and Justice and Home Affairs, which are outside the Community and Commission framework.

Struggling with Over-Ambition

The Union set very ambitious goals for itself in 1996, and had hoped to finish its work relatively quickly. Three months into 1997, however, many of the most controversial issues, particularly those dealing with majority voting and relative voting power, had not been resolved. Indeed, some, including Chancellor Kohl, had even suggested there might be a need for a 'Maastricht III'. The Union, however, had implicitly committed itself at the 1995 Madrid Council meeting to begin negotiations with potential new members six months after the completion of the current IGC. There was continued controversy over the desirability of beginning such negotiations simultaneously with all who had formally applied for membership

(Bulgaria, Cyprus, the Czech Republic, Estonia, Hungary, Latvia, Lithuania, Poland, Romania, Slovakia, Slovenia and maybe even Turkey), or whether to begin with a smaller group.

Nor was there agreement on the timescale for accession. While Chancellor Kohl had spoken, in 1995, of an agreement on Polish membership by 2000 and President Chirac, in December 1996, spoke of Hungarian membership by 2000, the Commission was more sceptical of this time plan. In a controversial January 1997 fact-sheet, the Commission maintained that accession could not be effected any earlier than 2002. Yet even the 2002 date was beginning to seem overly ambitious in light of Europe's simultaneous effort to achieve monetary union, revise the Maastricht Treaty, and begin NATO enlargement before the turn of the century. Europe may have to be content with reaching one of its goals by then; turning all the plans into reality simultaneously looks increasingly out of reach.

Reforming and Enlarging NATO

The wars of succession in the former Yugoslavia provided the primary catalyst for the post-Cold War process of reform currently under way within NATO. Given its apparent health and vitality today, it is salutary to remember that only in 1993 NATO faced a major internal crisis as the United States and its European allies disagreed fundamentally over how to resolve the conflict in Bosnia-Herzegovina. The US preference for a 'lift-and-strike' policy in the face of UK and French objections threatened not only to undermine transatlantic solidarity, but even placed NATO's very existence in jeopardy. It was only at this point that the Alliance, faced with a major existential crisis, finally began to establish a concrete programme for reform. This programme, forcefully promoted by the first Clinton administration in Washington, was unveiled at a summit meeting in Brussels in January 1994.

Two key sets of policies outlined at this summit form the main focus of the reforms. The first was the expectation that the organisation would plan to adopt new members from Eastern and Central Europe while establishing a Partnership for Peace (PFP) programme in the interim as immediate and concrete evidence of its willingness to cooperate with non-NATO European states. The second was the consolidation of an internal reform process to make the Alliance more flexible and potentially more European in orientation. This would be achieved through the Combined Joint Task Forces concept. Since 1994, implementing these two sets of proposals has dominated NATO policy-making and has been its central focus in 1996–97. At its summit in Madrid in July 1997, the Alliance is expected to select the first candidates to begin negotiations for membership, and to announce details of its internal reform and restructuring.

Keeping the Peace in Bosnia

Developments in the former Yugoslavia since the 1994 Brussels summit have continued to influence and provide the momentum for these policy changes. During 1995, NATO's more interventionist posture, which in August–September 1995 included extensive aerial bombardments of Serb installations, was seen as contributing to the resolution of the conflict and the signing of the General Framework Agreement for Peace in Bosnia and Herzegovina (the Dayton Accords) in December 1995. Whether NATO's intervention was the most significant factor in prompting the Agreement, or whether battle fatigue was a more decisive element, is open to argument. For NATO, the key gain was that its reputation was immeasurably enhanced. By the beginning of 1996, it had forcefully reasserted its credentials as the only military organisation in Europe able to deal effectively with the 'hard' security demands of the post-Cold War era, while delegating the 'soft' security elements of crisis-management to bodies like the Organisation for Security and Cooperation in Europe (OSCE), the European Union and the UN.

The NATO-led Implementation Force (IFOR), established in December 1995 to implement the Dayton Accords and take over from UNPROFOR, also helped to consolidate the Alliance's reputation. In terms of its military purpose, IFOR was a considerable success. The Bosnian cease-fire held throughout 1996, there was no resumption of hostilities, and no significant challenge to IFOR's mandate. More importantly, particularly for sensitive Western publics, it suffered no casualties in combat. The coalition, involving 33 NATO and non-NATO countries and over 60,000 personnel, has also been presented as concrete evidence of the Alliance's successful adaptation to the need for flexibility to permit cooperation with non-NATO partners. This is particularly important in dealing with the type of non-traditional crisis-management operations increasingly required in the post-Cold War security environment.

IFOR, however, did meet with criticism, in particular for its reluctance to engage more fully in the civilian aspects of the Dayton peace process and for not forcefully apprehending the notorious indicted war criminals living in the territory under its control. Yet, these criticisms have done little to tarnish NATO's record in the former Yugoslavia. The very fact that the media coverage of developments in Bosnia-Herzegovina during 1996 has been so limited has itself been seen as confirmation of IFOR's success. In December 1996, the smaller Stabilisation Force of about 25,000 soldiers took over from IFOR for an 18-month mission during which it will progressively reduce its presence to a deterrent posture and eventually withdraw completely.

From NATO's perspective, the main benefits of the Dayton Accords and IFOR's success in implementing them is that the issue of Yugoslavia has ceased to undermine the Alliance. Instead, developments in the Balkans have helped transform NATO from an organisation which was questioning its very reason for being, to one with renewed vigour, vitality and sense of mission. As such, the Yugoslav conflict has been the catalyst both for initiating NATO's reform process, and for its forceful and rapid implementation. The relative speed and progress with which NATO has resolved its internal problems and promoted eastward enlargement is in striking contrast to the parallel reform process under way within the European Union.

Cutting the Fat

Although the issue of enlargement has dominated media coverage and public perceptions of NATO's reform programme, eastward expansion is only one dimension of a dual-track process. The need for radical internal reforms within the organisation's existing structures, not least to adapt them to the accession of new members, has been just as critical a challenge for NATO leaders and officials. Such reforms are, in a sense, pre-conditions for enlargement. They have three main aspects which, although interconnected, have distinct features:

- to secure a new, more streamlined command structure;
- to implement and provide substance to the CJTF concept;
- to provide the necessary arrangements for a European Security and Defence Identity (ESDI) within NATO, which would also provide incentives for Spain and France to join its Integrated Military Structure.

Reforming the command structure is driven by a number of practical considerations. The reduced national defence budgets that reflect member-states' efforts to reap the benefits of the post-1989 peace dividends makes maintaining NATO's 67 Cold War major commands unsustainable. In addition, as Alliance strategy has shifted from defence against a monolithic Soviet threat to security against 'multi-faceted and multi-directional' risks, there is clearly no logic in concentrating commands on the northern European plains. Thus, the benefits of decentralising and regionalising the command structure, which could be seen as NATO's variant on the EU's subsidiarity principle, has been recognised.

However, reducing the number of commands to 35–40 has created an intensely political inter-governmental bargaining process where national pride can easily be stirred. In a process where there will inevitably be more losers than winners, the stakes are even higher. Although much of the bargaining has been shrouded in secrecy, the publicity in 1996 surrounding the divisive Franco-US dispute over who should lead NATO's Southern Command gives some indication of the sensitivities involved. While France, supported by Germany, has argued that a European officer should have overall control to reflect the ESDI and the Europeanisation of

NATO, the United States has insisted that leadership of Southern Command must remain a US prerogative, not least because it has the most military forces in the region, including the US 6th Fleet. As long as this dispute, and potentially others like it, remains unresolved, the prospect that reforms will be in place by the Madrid summit in July 1997 must be in doubt.

Table 1 *Western European Union Defence Expenditure, 1985, 1992–1996*

Constant 1996 US$m

	1985	1992	1993	1994	1995	1996	Change 1985–96 %	Change 1992–96 %
Belgium	5,735	4,466	3,972	4,096	4,539	4,282	-25.3	-4.1
Denmark	2,913	3,068	2,844	2,823	3,181	3,120	7.1	1.7
France	45,508	48,781	45,161	46,100	48,736	46,647	2.5	-4.4
Germany	49,125	45,366	39,435	37,728	41,991	39,105	-20.4	-13.8
Greece	3,245	4,738	4,315	4,506	5,159	5,620	73.2	18.6
Luxembourg	89	124	115	131	145	135	52.4	9.4
Netherlands	8,286	8,545	7,481	7,412	8,174	7,845	-5.3	-8.2
Portugal	1,708	2,738	2,324	2,257	2,724	2,908	70.3	6.2
Spain	10,497	9,797	8,790	7,711	8,827	8,729	-16.8	-10.9
UK	44,418	46,842	37,178	36,127	34,084	32,439	-27.0	-30.7
Total	**171,524**	**174,465**	**151,614**	**148,891**	**157,562**	**150,830**	**-12.1**	**-13.5**

Source: IISS, London

The second strand of the internal reform process has been to implement the Combined Joint Task Forces concept. In essence, CJTF represents a doctrinal and organisational rationalisation of the *ad hoc* alliances forged in such conflicts as the 1991 Gulf War and the Yugoslav conflict. CJTF is NATO's formal answer to using assets from its Integrated Military Structure for a wide range of potential collective security operations which might include both NATO and non-NATO members. IFOR is a clear example of a CJTF in practice. Implementing the concept is also closely connected to reforming the command structure, since at least some, if not all, of the new commands will have to accommodate potential CJTFs as well as the traditional Article 5 collective defence provisions.

However, the concept has also contributed to the third element of the reform process – the strengthening of European defence capabilities within the Alliance to create an ESDI without undermining NATO's strategic primacy. The CJTF has been the principal mechanism for Europeanisation.

In early 1996, there were intensive negotiations over how a European-only WEU-led CJTF could be allowed to use NATO assets. Again, the negotiations tended to pit the US against France, with the former insisting upon, and the latter resisting, a US veto of any such European-led operation.

In the end, a compromise was announced at the Berlin ministerial meeting in June 1996. In effect, the United States secured the right to approve any such European-led operation, thus also permitting Washington to veto any mission it opposed. As a consequence, French claims that 'for the first time in Alliance history, Europe will be able to express its personality' must be taken with a pinch of salt. Even if Europe did seek to pursue an independent operation without US participation, itself rather unlikely, the key assets required are not NATO assets, but US-owned assets, such as satellite intelligence systems and heavy-transport air-lift capabilities. In reality, the large-scale post-Cold War defence cuts in all European states are undermining the prospects for a viable European defence capability and increasing rather than decreasing Europe's dependence on the United States.

France's prominence in pushing for the greater Europeanisation of NATO reflects both its traditionally uneasy relationship with the Alliance, and its particular difficulties in adapting to the changing security environment after the end of the Cold War. Since the collapse of the Warsaw Pact, the French government has had to reassess its traditional defence posture. This was based on its independent nuclear deterrent and on forging a European-only defence capability, principally through the WEU, to overcome perceived US hegemony within NATO. The nuclear deterrent is steadily being de-emphasised, and France has come to realise that a European defence identity distinct from, and not subsumed within, NATO is not practical, not least because of the cuts in the defence budgets of European states, including France. These developments left Paris with little choice but to seek accommodation with NATO and to pursue greater military cooperation with the United States. Europeanisation of the Alliance, even if essentially a façade, gives France a justification for gradually integrating into NATO's military structures, thus overcoming the Gaullist legacy. France's decision in December 1995 to participate fully in NATO's Military Committee, and the fact that French ground troops in IFOR were under direct NATO command for the first time since 1966, are important landmarks in this re-orientation of French defence policy.

For the other allies, most notably the US, the UK and Germany, the quest to Europeanise NATO has the indirect benefit – much like the promotion of the WEU in the past – of potentially securing a greater French commitment to NATO. In addition, it helps to end the Spanish and French 'opt-outs' from the Integrated Military Structure, thus preventing potential new members of the Alliance from being tempted to pursue similar partial membership. In November 1996, Spain agreed to join the

Alliance's new structure. France, however, continued to hold back, arguing that it would only rejoin if and when NATO became sufficiently Europeanised. As with France's reactions to economic globalisation, the drive towards NATO integration touches a sensitive nationalist nerve. Over issues like the leadership of Southern Command, Paris demonstrated its almost unique capability – or even need – to irritate the Americans. The rows in Paris following the publication of a Franco-German defence paper in January 1997, which firmly identified NATO as the guarantor of Europe's strategic defence, also highlighted the domestic constraints inhibiting President Jacques Chirac from dismantling the Gaullist legacy, with ending conscription in 1996 the most notable development. The possibility that Gaullist pique will continue to exert its obstructionist influence represents a significant underlying threat to the Madrid summit in July 1997.

Spreading Eastward

While internal reform decisions will be important at the Madrid summit, the dominant issue will undoubtedly be that of NATO enlargement. During 1996, the major development in the intellectual debate over enlargement has been a shift from the vigorous, if not always enlightening, arguments for and against to the acceptance that expansion will almost inevitably take place. The debate has thus progressed to how the process can best be managed to minimise any potentially negative repercussions, particularly in relation to Russia and those left outside the Alliance after the first wave of enlargement. Two developments have underpinned this shift in focus. In September 1995, NATO published its 'Study on NATO Enlargement', which set out the reasons for its commitment to eastward expansion and the broad outlines of how this was to be promoted. Then, at the ministerial meeting in Brussels in December 1996, NATO governments announced that the first invitations for new members to join would be issued at the Madrid summit in July 1997. Even the most sceptical commentators recognise that some Central and East European countries will almost certainly be invited to open negotiations to join as full members.

The pragmatic issue of how to manage enlargement has now become the primary challenge facing the Atlantic Alliance. There are three main aspects to this challenge. The first is perhaps the easiest to resolve: which country or countries should be invited to be the first new member-states? During 1996, 15 East and Central European states engaged in 'intensified dialogues' with NATO to assess their suitability as prospective full members. These dialogues were primarily technical in nature, dealing with issues such as budgeting, command structure, the presence of foreign troops, the prepositioning of equipment and integration into the force-planning process. The actual decision of which countries to select will, of course, be a political one, and the December 1996 Brussels meeting cautiously only committed itself to inviting 'one or more countries'. This cryptic declaration could not hide the general recognition that the front-runners were Poland, Hungary and the Czech Republic, with Slovenia and Romania as potential Balkan entrants. These countries are widely considered to be among the most advanced in terms of political and economic reforms.

The Madrid summit will only issue invitations and not full membership to the prospective members. This will be followed by further bilateral negotiations to be completed by 1999 in time for NATO's 50th anniversary. Some potentially dangerous pitfalls, however, still remain. The accession of a new member into the Alliance needs to be ratified in all 16 member parliaments. In late 1996, Turkey stated that it might use its power of veto unless the EU looks more favourably on its long-standing claim to join the Union. Similarly, Denmark has hinted that it might derail enlargement unless some or all of the Baltic states gain compensation for their exclusion from NATO by fast-track incorporation into the EU. Whether the US Senate will continue to maintain its bipartisan support for enlargement is also in doubt, particularly once the financial costs are made known and fears of new military commitments are raised.

The selection of new members is less problematic than the second major challenge – what to do with Russia. There is a formidable consensus in Russia, including within even the most liberal groups, that NATO enlargement represents a threat to Russia's core national-security interests. The West's argument that enlargement would consolidate stability in Central Europe, as much in Russia's interest as in the West's, persuades no one in Moscow. To Russians, NATO is the most powerful military alliance in Europe and has demonstrated, as when it bombed Serb installations in Bosnia in September 1995, that it has the will both to ignore Russian interests and to engage in aggressive out-of-area military actions. In addition, the open-endedness of the enlargement process, which makes countries like Ukraine and the Baltic states potential future candidates, feeds Russian fears that the West is pursuing a rolling containment policy.

There is much bargaining to be done. The Russian politico-military establishment is well schooled in the art of *realpolitik* and realises that, since enlargement is going to take place, Russia may be able to negotiate a deal that could lessen some of its perceived damaging consequences. Thus, while Russia's opposition to the principle of enlargement remains firmly non-negotiable, Moscow has indicated that it is not implacably opposed to a NATO–Russia agreement.

Although this sounds positive, a large problem remains. The two major concessions Moscow seeks are simply not politically acceptable to NATO governments. Russia is seeking an institutionalised 'voice' within the Alliance which would give it the *de facto* right to veto any developments that it opposed. It is also demanding treaty-bound legal commitments on issues like the non-deployment of nuclear weapons and non-stationing of foreign troops on the territory of any new members, since it does not trust the West to honour its promises on these matters.

For its part, NATO is willing to provide any number of fora or institutions to facilitate dialogue, cooperation and joint action with Russia, but the Alliance will not accept any derogation of its independence of decision-making. Western governments have also proposed various forms that a NATO–Russia agreement might take, such as a NATO–Russia Charter, a joint NATO–Russian Council, a joint brigade, and a review of the 1992 Conventional Armed Forces in Europe (CFE) Treaty, but all of these have fallen short of the legally binding treaty that Moscow demands. On issues like the deployment of nuclear weapons and the stationing of foreign troops, NATO ministers stated at their December 1996 meeting that they 'had no intention, no plan, and no reason to deploy nuclear weapons on the territory of new members ... and do not foresee any future need to do so', but refused to rule out the possibility of a future shift in policy. NATO governments in general, and the United States in particular, are determined not to tie their hands over any agreement with Moscow, particularly when the future evolution of Russia is itself so unpredictable.

Even if a NATO–Russia agreement is finalised in time for the Madrid summit, it is unlikely to satisfy fully Russia's minimal demands. Instead, Moscow is likely to seek to impose some costs on enlargement. At the very least, it is likely to revise its defence posture, probably by placing greater emphasis on nuclear deterrence to compensate for NATO's perceived conventional superiority. Russia also has a number of other potential means through which to express its opposition: it could withdraw from SFOR and the PFP programme; it could reduce its commitment to the OSCE and force an OSCE withdrawal from Chechnya and the Caucasus; it could be less amenable to CFE adaptation and refuse to ratify the Strategic Arms Reduction Talks (START II) Treaty; it could seek to strengthen the Commonwealth of Independent States (CIS), placing greater pressure on Ukraine and seeking to consolidate the CIS defence and security provisions; and it could generally be less cooperative with the West over its policies towards non-European regions such as China, East Asia, Central Asia and the Middle East. Whether Russia will actually pursue any of these options will ultimately be determined by the extent to which the costs of a more confrontational relationship with the West override the perceived benefits of expressing its dissatisfaction over NATO enlargement so forcefully.

The difficulty of forming a satisfactory NATO–Russia relationship greatly complicates NATO's third major challenge – what to do with those Central and East European countries not included in the first tranche of enlargement and who may have to face a dissatisfied and more hostile Russia. Belarus has resolved this potential problem by folding itself into a Russian sphere of interest. But this approach does not attract the other countries in the region and the dangers are particularly great for Ukraine and the Baltic states, and perhaps to a lesser extent for Slovenia and Romania. The problem facing NATO leaders is that by increasing the security of those countries granted full membership, the sense of insecurity of countries like Ukraine and the Baltic states will inevitably become more intense. Given that these countries remain unstable politically and economically and have difficult bilateral relations with Russia, the West must take measures to enhance their stability and, at the same time, compensate them for not being offered the prospect of immediate membership.

Resolving these difficulties depends on Russia's response to enlargement and the details of the NATO–Russia agreement. NATO has promised Ukraine a formalised NATO–Ukraine relationship, following the Russian model. Apart from further expressions of the West's commitment to Ukraine's territorial integrity and national sovereignty, NATO will seek to reinvigorate the PFP programme so that, in the words of US Ambassador to NATO Robert Hunter, 'we make the difference between being a partner and being an Ally razor thin'. Indeed, to assuage the fears of those PFP countries unlikely to be included in the first wave of enlargement, NATO has created a 'Senior Level Group' that meets regularly in Brussels to discuss the contents of an 'enhanced' PFP programme. Such a programme could involve mechanisms for greater political consultation and, possibly, for participation in non-Article 5 missions, including CJTF planning. Clearly, progress can be made in this direction but, as is widely recognised, the strategic prize of Article 5 is the ultimate guarantor of stability. This is what PFP, however it is institutionalised and consolidated, fails to provide.

A more profitable avenue would be to engage the EU far more intensively in the economic and political developments of the former Warsaw Pact East European states. In reality, a new economic iron curtain has been created which rigidly separates the reforming countries of Central and Eastern Europe, which now enjoy practically unrestricted free trade with the EU, from those countries which do not. The EU's material support for Ukraine is also paltry; of the $5bn Ukraine received in 1995, the net EU contribution was $30m. There is clearly a strong case, which Russia is itself explicitly demanding as a *quid pro quo* for a NATO–Russia agreement, for the EU to extend its free-trade agreements further East and play a much more prominent role in cementing economic and political stability in the European states of the former Soviet Union. In general, the need to provide greater links between the supposedly parallel processes of NATO and EU enlargement is essential and will undoubtedly become a more salient issue after the first new members are invited to join NATO.

The Need for Clarity

The difficult strategic dilemma which Ukraine will continue to face after the Madrid summit in July 1997 highlights the critical importance of successfully managing the NATO reform process in general, and eastward enlargement in particular. The decisions made in Madrid will be significant not only in Europe, but also further afield. For Europe, the summit will give a clear indication of whether Western Europe and its most successful institutions are truly committed to a pan-European mission and whether there is a genuine determination to overcome the divisions of the Cold War. For the United States, the success or failure of eastward enlargement, which President Bill Clinton has made the centrepiece of his personal contribution to a new post-Cold War security framework, will have a major impact on future US commitment to a European, and broader global, security role. And, clearly, Russia's response to NATO enlargement will be critical in determining its future willingness to cooperate with the West.

It has frequently been said that NATO's function was to ensure the involvement of the US in Europe and reintegrate Germany within the political European world, while defending against a potential threat from Russia. To a significant extent, with the exception of the last measure which the end of the Cold War has made moot, this reasoning continues to hold true for NATO's reform programme. As early as 1993, US leaders were stating that the Alliance must 'enlarge or die', and a restructured and enlarged NATO has been an implicit condition for securing continued US commitment to European security. The German government has also been a major force behind enlargement, primarily perceiving it as a mechanism for keeping itself 'down' so that its inevitable intensified engagement in Central and Eastern Europe should be promoted through multilateral and institutional channels rather than independently and unilaterally. And, for all NATO's willingness to countenance much closer cooperation and joint action with Russia, in the final analysis Moscow is being excluded from, rather than being invited into, the inner sanctum of the Atlantic Alliance.

The critical question is whether these elements of continuity in NATO's core mission represent an appropriate and sufficiently imaginative response to the dramatic events of 1989 and 1991 when Europe was reunified and the Soviet empire collapsed. The importance of the Madrid summit in July 1997 to provide clear answers to this question cannot be overstated. Even so, the summit meeting is only the first step in a new and delicate process of negotiations and bargaining, both bilateral and multilateral. Indeed, before new members can take up their permanent seats in the North Atlantic Council, a number of hurdles must be overcome. Chief among these is the ratification process that will have to take place in each of NATO's 16 countries, whose outcome is by no means a foregone conclusion. Once this obstacle has been overcome, command relations, subject to review for some time, will also have to be re-examined in light of an expanded membership. Equally important, however, will be the need to give some substance to an 'enhanced' Partnership for Peace programme for those countries not invited to join NATO in Madrid. Finally, and perhaps most critically, the content of a proposed 'Charter' or agreement to define the NATO–Russia relationship has also to be worked out, as does the issue of whether a separate agreement should be reached with Ukraine. There is plenty here to keep the Alliance busy in the two years that remain before it celebrates the 50th year of its existence.

Russia: Still Searching For Stability

In writing about Russia, it has become a convention to allude at some point to the 'Time of Troubles', those periods that coincide with a loss of leadership and a corresponding fall into chaos, anarchy and poverty – until another iron fist seizes Russian power, restores order with brutality, and drives economic advance by the knout. Although there were problems in Russia in 1996 and early 1997, signs that its political and economic reforms were irreversible became clearer. In 1996, the country survived one of the most politically challenging periods in its five-year history. This included the first truly competitive presidential elections, the first all-Russian local elections, the end of the war in Chechnya, the crisis of state power following President Boris Yeltsin's absence due to ill-health for much of the year and, finally, the first year of a majority pro-communist parliament working with a reform government on crucial economic and political issues, including approving the new state budget, implementing military reform, and wrestling with the non-payment of wages and other economic problems. Any one of these challenges could have reduced Russia to chaos, but it struggled through, ending the year with a surprisingly strong economy, a functioning (if often challenged) system of checks and balances and no significant deviations from its course of reforms.

These two parallel trends – economic progress on the one hand and continuing crisis on the other – have created a climate of uncertainty in post-Soviet Russia that is unlikely to change in the next few years. The major problems are:

- a continuing decline in industrial production which threatens to increase unemployment and social unrest;
- a deteriorating military faced with a declining budget and significant personnel cuts; and
- a profound criminalisation of the country's political and economic spheres that hinders its transformation to a more democratic and market-oriented society.

But this is not a time of troubles in the classic sense. Russia may become, and is already becoming, less democratic and more arbitrary in its rule, but there is no tyrant waiting in the wings, nor could one impose himself on a population that is now half-free. Nor is it, as the optimists claim, merely a time of transition from something nasty to soon-to-be-achieved stability. Although this is more likely than descent into fully-fledged authoritarianism, there is no guarantee.

If this period needs a label, it could be the 'Time of the Trough'. Russia is a country whose profound economic and political crises have yet to be fully faced and internalised; whose ruling élite is still consolidating, and enriching, itself; and whose very borders are still in question. The shock of the Soviet Union's collapse in 1991 is still being absorbed, and will continue to be for some years. What is clear is a growing consensus among Russia's political élite, across almost all parties, that stability is the major priority. The theme of stability will continue to dominate domestic economic growth and even a struggle for power among the élite. In 1996, events indicated that this stability might be sustained even without the strong presence of President Yeltsin, previously seen as the primary guarantor of Russia's reforms.

Grim, But Not Desperate

In most areas of Russian life the progress made in the early years of reform has either proved illusory, has ceased or is now compromised. Government continues, and basic services, if increasingly basic, operate. Moscow at least is a booming city. Curiously, the economy, often seen as the most disastrous aspect of the post-Soviet period, showed the greatest signs of hope in 1996. The relatively tight macroeconomic framework in place for much of the past three years was maintained. Inflation remained low by Russia's past standards, dropping to 22% in 1996 – its lowest point since 1992 – from 131% in 1995. Pressure from the main lobby groups to increase spending has been contained, if at times with difficulty. Foreign trade was a bright spot in 1996, with exports continuing to rise, resulting in a provisional trade surplus of $28bn, up from $21bn in 1995.

But black spots remain, even if the 6% drop in production has reached a plateau. Tax collection was still seriously low in 1996, reflecting both the inexperience of the tax-collection service, and the extensive tax concessions granted by ministers and the President himself to companies run by friends and associates. Banks and finance houses – which have benefited enormously from the second stage of privatisation when energy and other assets were sold at rock-bottom prices to chosen enterprises – have acquired companies which have not yet been restructured, despite new managements. Opposition from the workers, lower management and local authorities has stalled far-reaching reform in all but a handful of companies, leaving the major reforms still to be tackled.

The business élite remains composed of those who generally occupied the lower rungs of the Soviet state or Party ladder, or who were outside it altogether. Their attitude in the last years of the Soviet Union and the first years of Russia was one of almost exaggerated business individualism. Now most of them work with and through the state, having learned that it is the state that hands out the privileges and the assets they all need. The slow rate of liberalisation, the limited role played by foreign capital, the isolation of the currency and lax standards all slow the integration of this élite into the world, shielding them from the need to adopt practices which would elevate their business culture.

Until that is done, their products and practices will remain backward, even as their profits, from trading and money manipulation, remain large. The financial markets are slowly being liberalised and becoming more sophisticated. The role of foreign companies and investors, largely in portfolio investment, is growing. But Russia remains low down the list of emerging markets in which funds or individuals are prepared to invest, and much of the $20–30bn in foreign bank accounts remains abroad, except for 'hot' money returning to take advantage of high interest rates.

The State of the Nation

Russia is a 'quasi democracy' with all the elements of a legitimate, constitutional state: a Constitution, adopted at the end of 1993; an upper and lower house, elected by universal suffrage; a hierarchy of courts, including a constitutional court; elected local and city assemblies; independent, as well as state-owned, newspapers and television channels protected by law; and a dense network of commitments, some international, to civil rights.

The 'quasi' nature of Russian democracy lies not in the letter, but in the practice, of its laws. The Constitution is heavily presidential, and while the two houses of parliament (the Duma, or lower house, and the Federation Council, or upper house) have powers to dismiss, call to account and ultimately impeach the President, the Constitution also allows what amounts to rule by presidential decree, and the mechanisms by which parliament can effectively control the executive and the presidency are cumbersome and slow.

This 'quasi' democracy became more apparent during the Russian presidential race that dominated the first half of 1996. Although the communist opposition, headed by Gennady Zyuganov, had begun its presidential campaign in mid-1995, Yeltsin's supporters only began actively campaigning to re-elect him in February 1996. Until January 1996, Yeltsin himself seemed uncertain whether or not to stand because of his poor health. Unlike many other former Soviet republics, Russia has not yet experienced the transfer of power from one post-Soviet President to another. Although a backlash towards neo-communist leaders has been evident in many Eastern European states, it could have a devastating effect in Russia. Given its current political environment, a combination of a neo-communist President with almost dictatorial powers and a communist-dominated parliament could have set Russia back many years in its reform efforts. Of the reformist candidates, only Yeltsin seemed capable of winning over unified communist forces. Thus he faced serious pressure to run for office again, although his chances in early 1996 did not look good – according to some opinion polls, he had no more than 15% support and carried with him heavy political baggage including serious economic problems and the extremely unpopular war in Chechnya.

But Yeltsin is a 'master of come-backs' and did not compromise his reputation. Instead, he returned as a strong, assertive leader, able to distance himself from those perceived to be responsible for delaying wage payments, and he succeeded in blaming the country's economic hardships on the regional and local levels where federal funds were often stolen or invested in private bank accounts to generate interest. Yeltsin campaigned actively, travelling extensively throughout Russia despite his health problems. This paid off, and in the first round of elections on 16 June he narrowly defeated his main rival, Zyuganov, winning over 34% of the votes to Zyuganov's 32%. The first round, however, brought unexpected success to another candidate – recently retired General Alexander Lebed – running for the first time with no strong party backing. He finished third with 14% of the votes.

The second round of elections was only contested by Yeltsin and Zyuganov, both of whom had won all the votes they could expect. The decisive factor was thus who could win Lebed's support. Yeltsin won by moving first, announcing almost immediately after the first round that Lebed was to head Russia's Security Council and be Yeltsin's personal National Security Adviser. Then Yeltsin strengthened state control over media coverage of the presidential race, giving the communists no opportunity to project their new image. Instead, they were portrayed as the harbingers of the economic hardships and political repression of the Soviet era. These strategies helped Yeltsin to a major victory with 53.8% of the votes compared to Zyuganov's 40.3%. But this overwhelming win cost Yeltsin dearly in terms of his health. A week before the run-offs, he suffered a second heart attack that removed him from the political scene for much of the rest of the year.

The overarching political reality of the rest of 1996 and the first part of 1997 was President Yeltsin's withdrawal from the political scene, and the concomitant impossibility of replacing him while he still lived. Although Yeltsin is often seen as a transitional figure, post-Soviet Russia could not have been led by one who marked too sharp a break with the communist past, as in the Czech Republic or Poland. Weakened by two heart attacks, however, Yeltsin found it more and more difficult to keep pace. On 5 November 1996 he underwent a quintuple heart by-pass operation from which his recovery was slowed by an attack of double pneumonia. By the end of 1996, Yeltsin was a shell of his past self, ailing, erratic, sustained by an entourage who owed their position and frequently their fortunes to his patronage. But by early March 1997, he had returned to Moscow, looking fitter and acting with determination. He sacked his cabinet and appointed two leading reformers, Anatoly Chubais and Boris Nemtsov – a young liberal provincial leader – as Deputy Prime Ministers. Chubais was also named Finance Minister, indicating Yeltsin's intention to revitalise the economic reforms.

Shifting Power Centres

The strength of Russia's political centre is mainly associated with the Moscow finance houses that handle some 80% of Russia's financial transactions. On the political level, however, the all-Russian gubernatorial elections in 1996 further legitimised the rapidly growing political independence of the outlying regions. The actual process of *de juro* federalisation had started in late 1993 with the adoption of the new Russian Constitution. However, it only gained real substance when Yeltsin, seeking support for his presidential campaign, authorised the practice of concluding special agreements on dividing power between the federal centre and individual regions. By the end of 1996, most of the Federation's entities had signed agreements with Moscow, each different in scope and substance and tailored to the entity's individual needs.

These agreements could only be developed and implemented if power was transferred to legitimate representatives of regional élites. Thus Yeltsin could no longer postpone elections for regional executives who had previously been directly appointed by the President. Without regional elections, the *de facto* political rise of the regions threatened to undermine Russia's territorial integrity. These elections returned a total of 60 governors between December 1995 and January 1997. Between September 1996 and January 1997, 48 regional governors were elected. According to an Open Media Research Institute Regional Report, 15 of these represented the Communist Party and its political allies; six were industrial managers who ran as strong independent candidates; two were elected on the strength of their close association with Lebed; while Vladimir Zhirinovsky's Liberal Democratic Party only managed to elect one.

Although the balance of political forces appeared strongly to favour the communists, several of the newly elected governors disassociated themselves from the Party immediately after the elections. They had used Communist Party support and funding to get elected, but were not prepared to carry out its ineffective economic policies in their regions. Similarly, many incumbents only used the strong backing (including financial) they received from presidential circles to gain power, defending their regional interests once they had assumed their seats in the Duma.

The power of these regional governors and politicians is growing in the absence of a clear federal settlement between the centre and the regions and because the government and President need allies to stay in power. While the new governors tend to be conservative, they are no longer interested in ideology. Instead, they are concerned with power and money, and in attracting to their region what investment they can. They lobby ferociously for state funds, and are strengthening alliances with local military and police commanders, with local business élites and with the heads of the collective farms and state institutes. The sign of a 'successful' governor is his ability to realise his region's financial and political potential, by raising revenue from trade, issuing licences, granting rights and from privatisation.

Governors, mayors and elected leaders face a crisis of staggering proportions: the continuing cuts in central government expenditure and the weakness of the regions' own tax bases means that the infrastructure they have to maintain, as well as what they are unwillingly inheriting from other enterprises which are shedding their responsibilities – like nursery care, medical clinics, cultural centres and housing – is now in very bad shape. Trains and buses are not replaced, houses are not repaired, sewers fall in, the water supply deteriorates, and no new facilities are built as capital and maintenance budgets are channelled into paying wages and maintaining basic necessities.

A More Sinister Shift in Power

Insinuating itself in every area of public, and even private, life is crime and corruption, a phenomenon which has changed from being marginal, if dramatic, to being integral to the life of the country. Its dramatic side is organised crime, which levies a tax on enterprises estimated at around 20% of their total turnover. These groups took their cut from oil and gas lines, from the trade of most commodities and from the proceeds of privatisation. They controlled the booming drug trade from Central Asia and the Caucasus through Russia to Europe and the US. They ran guns to and from unstable areas, created international prostitution rings, bribed politicians and intimidated real or potential enemies.

Yet, bad as this has been, it is arguably the less serious part of the problem. More serious is the grey area of corruption, of 'understandings' between officials and business people who pay the former large sums to protect their interests within the bureaucracy. The increasing closeness of the different clans – finance, energy, industrial – enables them to divide the economy between them. They usually have greater interest in keeping inconvenient change at bay than in increasing economic efficiency. They only allow foreigners access on particular terms, at times demanding large bribes, rejecting their full participation for fear they would take their profits out of the country.

A Disintegrating Military

Crime has corrupted the top echelons of the military command, as well as other groups. Publicised scandals involving generals and admirals no doubt mask a much larger iceberg of corruption which has managed to remain hidden. The rottenness of the military began at the head, but now permeates the body. Even though its efficiency was overestimated in the Soviet period, the military was clearly effective in some areas.

The replacement of Defence Minister General Pavel Grachev by General Igor Rodionov on 18 June 1996, immediately after Yeltsin's re-election, was advertised as heralding a new era, including an attack on corruption. Rodionov was seen as the Minister of military reform. His predecessor had delayed implementing such reform for fear of losing power not only over the military, but also in government where his position was very weak. Rodionov had the strong backing of Lebed and later of Yeltsin himself, and enjoyed the support of parliament and stable relations with the military. But not even this political capital was enough to initiate reform, and Rodionov was soon reduced to desperation, prepared to use all available means to secure basic funding to feed his soldiers. By the end of 1996, he described himself as 'Minister of a disintegrating army and a dying navy'. In a bid to draw attention to the military's plight, he even referred to the threat posed by the lack of funding to the reliability of Russia's nuclear weapons. This did nothing to improve the process of military reform, and instead undermined Rodionov's popularity, exacerbating the crisis within the Ministry of Defence itself.

Despite Rodionov's campaign to increase funding for the military, the Ministry of Defence received only about 70% of the funding it was allocated in the 1996 state budget. There was widespread hunger and even death among conscripts, especially in the remoter military districts. According to Ministry of Defence estimates, in 1996 the armed forces only received about 39% of their necessary food requirements and only 2% of their actual healthcare needs. Many military personnel took other jobs to support their families, because salaries were often delayed for months – in some regions for up to five months.

The combat readiness of the armed forces has fallen to its lowest ever point. Only about 30% of the 1996 budget for research and development (R&D) and weapons procurement was actually forthcoming during the year. According to Ministry of Defence estimates, only 15% of Russian communications and electronic warfare materiel and about 4% of helicopters, mechanised infantry combat vehicles and armoured paratroop personnel carriers meet world standards. Much equipment has simply rusted away, and some hardware has been sold by local commanders to criminal groups.

This financial crisis in the military occurred as Russia's arms sales rose. In 1996, Russia became the world's second leading arms exporter, receiving over $3.5bn from sales compared to $3.1bn in 1995. Russia placed special emphasis on increased sales to Asia, Latin America and the Middle East, and by March 1997, it had concluded several large-scale military cooperation agreements and sales to countries such as China, India and South Korea

Although lack of funding jeopardised Rodionov's plans for military reform and significantly undermined his popularity, he did become the first civilian Minister of Defence in recent Russian history. In December 1996, Yeltsin signed a decree discharging Rodionov from the military because he had reached retirement age, but retaining him as Minister of Defence in a civilian capacity. This did not, however, significantly increase civilian control of the military. The 'civilianisation' of the Minister of Defence, welcomed by the Russian government as well as by the military, remained only symbolic and gave Rodionov no more power to develop and implement military reform. Instead, other institutions assumed increased responsibility for Russia's security policies, creating chaos in relations between them.

One of these powerful new bodies was the Defence Council, created in October 1996 to counterbalance the increasing powers of Lebed who, while Head of the Security Council from June to October, wanted to control all defence and security policy-making. During 1996, the Council's power increased tremendously, while the Ministry of Defence's role continued to decline. In addition to overseeing military appointments, the Defence Council was responsible for drafting an official paper on military reform and new Russian military doctrine, actively participating in foreign-policy issues (including NATO enlargement) and drafting a military budget.

The military is also concerned to minimise as far as possible the effects of its greatest humiliation – the war in Chechnya. The defeat in Chechnya is a defining moment in post-Soviet Russian history, and its effects will be deeper and less predictable than is yet apparent.

Ending the War in Chechnya

The war in Chechnya began in December 1994 when Russian troops were sent in to the republic to end the defiance of then President Dzhokar Dudayev, who since late 1991 had declared Chechnya a separate state, making it a safe haven for gun and drug running. Moscow had tried to undermine Dudayev using the secret services to support his opponents. When that failed, it sent in an army whose commander, Grachev, promised to clean the place up in a matter of days.

Two years later, an army little more than a rabble engaged in a desperate struggle with men whose warrior traditions and tenaciousness made them impossible foes – at least for an army scraped together from ill-trained conscripts and former criminals, under-equipped, badly officered and with little idea of how to defeat their opponents. By the end of April

1996 these troops, using massive firepower from the air and tanks on the ground – causing many civilian casualties – had driven the Chechen fighters back to their strongholds in the mountains to the south. Just before, however, after a ferocious aerial bombardment of two of these mountain villages, Chechen troops showed how dangerous they still were.

On 5 March, six federal checkpoints in the Chechen capital Grozny were attacked and two regional police divisions seized. At least 2,000 Chechen troops had taken control of most of the city, and despite the efforts of vastly superior numbers of Russian troops the rebels held through two days of bitter battles. On the morning of 8 March, having demonstrated their strength, the Chechens faded out of the city. It was a devastating humiliation for the Russian Army and for Grachev, and led to a renewed effort by Yeltsin to negotiate a way out of the quagmire.

His plan, announced on 31 March, called for the immediate end of all troop operations in Chechnya, and negotiations with Chechen leader Dudayev through mediaries to arrange a treaty deliminating powers between the two entities. The cease-fire never came into effect and, in a bombing raid on 21 April, Dudayev was killed. Rather than halting the peace talks, this seemed to provide a new impetus. Yeltsin had vowed publicly never to talk directly to the arch-rebel Dudayev; his death opened the way for Yeltsin, in the midst of a desperate attempt to win a difficult election for another term as President, to meet Dudayev's successor, Zelimkhan Yandarbiev, on 28 May. While the leader of the rebel government was in Moscow, Yeltsin flew to Chechnya, met with the generals and servicemen there, and proclaimed a victory in the war.

It was a shrewd move. Yeltsin was now able to campaign for re-election on the basis that he had ended the Chechen war. While the rebel government achieved official recognition, what was not achieved was a cease-fire. The Russian military in Chechnya refused to recognise Yeltsin's agreement to a cease-fire from 1 June, continuing their ramshackle campaign with no more hopes of winning than ever. Once again the Chechens took things into their own hands. On 12 August, they launched another attack on Grozny. Three days later, as Yeltsin was being inaugurated after his presidential victory in Moscow, they completed their seizure of the capital. This time they had come to stay.

Yeltsin appointed Lebed as his special representative to reach a settlement, and then, at Lebed's demand, granted him effective authority over all Russian forces in Chechnya. Armed with these powers, Lebed's negotiations with Aslan Maskhadov, the Chechen commander in Grozny, were fruitful. On 3 September, a vague statement of principles, the 'Khasav–Yurt' agreement, was signed by the two negotiators. It put on hold the issue of relations between the Russian Federation and the Chechen Republic until 31 December 2001; it called for the establishment by 1 October 1996 of a joint commission to monitor a troop withdrawal, various financial and budgetary matters, prepare programmes to restore the Republic's socio-economic complex, and provide food and medicine to the population.

Vague though these provisions were, they were sufficient to bring about an effective cease-fire and to ensure the withdrawal of Russian forces. By January 1997, all Russian troops had been withdrawn, various commissions had been established, agreements had been reached on financial questions, and the ground-work had been laid for a presidential election in Chechnya. On 27 January 1997, the Chechens voted for three candidates, all of them commanders during the two-year war, all of whom advocated independence for the republic. Maskhadov won by a wide margin with 60% of the vote, more than double that of his nearest rival. Maskhadov was the candidate most committed to a continuing dialogue with Russia and to peaceful reconstruction. For these reasons, the Russian administration was relieved to see him the victor. Yet he insists that Chechnya is already an independent country, with only world recognition of this fact to be gained.

Eighteen months of Moscow trumpeting its victory ended with the realisation that a withdrawal had to be negotiated. The Russian Army has left, with the vague assurance that the republic will remain part of Russia for five years. President Maskhadov has inherited a *de facto* independent state, albeit a devastated one. Its capital – renamed Dzhokar-Ghala, in honour of the former president – is a ruin and its economy is at subsistence level. At stake is what kind of relationship it can negotiate with Russia, and how far Russian politicians will be able to treat Chechnya rationally, at least tacitly conceding its independence, or how far they will seek to punish it by other means.

The Struggle for the Presidency

While Yeltsin struggled to recover his health and position in 1996, there was much jockeying for power among possible successors. By early 1997, these were reckoned to be no more than four. The most reformist was Viktor Chernomyrdin, Prime Minister since the end of 1992. His main support came from the directors of state oil and gas enterprises, and members of the business élite. Gennady Zyuganov did well enough in the 1996 election to retain the leadership of Russia's Communist Party, but he could not shake the Soviet-era dust from his shoes. While still maintaining the support of pensioners and those nostalgic for the security of the communist era, he has lost considerable support within the old Communist Party itself. The candidate who did break the mould was Lebed. From summer 1996 onwards, he was consistently Russia's most popular politician, often by a long way. The authorities tried to ignore him, but by mid-January 1997 appeared to change tack, allowing him to appear on television, perhaps a sign of a coming *rapprochement* or even of an anointed candidate after Yeltsin. The wild card was Yuri Luzhkov, the rich and energetic mayor of Moscow. Whether his attempt to grasp the nationalist mantle will convince those in the rural areas of Russia that he is a man to back is still open to question. But he maintains a tight grip on Moscow's financial might, is close to the national media and has a considerable reputation for making things happen. Luzhkov has to be taken seriously. The candidate list revealed two other large changes. One notable omission was the liberal Grigory Yavlinsky, the other was the ultra-rightist Vladimir Zhirinovsky. Instead, the 'centre' was filled by those who played the patriotic card, were vague about economic reform, made unkeepable promises and emphasised their strength of will. It was not ideal, but neither was it fascist or communist.

Betwixt and Between

That so much symbolism was used to emphasise the past and future greatness of Russia pointed to the fact that none of the present contenders, nor anyone else, had a clear or popular view of Russia's place in the present. The 'Westernisers', who no longer trumpeted their positions as they had in the late 1980s and early 1990s, claimed that Russia's most important relationship had to be Europe. The nationalists either denied it, stressing that Russia was a place apart, or proposed that Russia had to find itself in the East as much as in the West.

The main source of contention was NATO expansion. Moscow's former Warsaw Pact allies clearly still feared Russia. Even some of its supposed allies in the Commonwealth of Independent States – especially Ukraine – looked enviously to the West, and found no vision of an alternative or opposing alliance to NATO.

Although Russia could bark at expansion it could not bite, and this realisation was all the more bitter as the message sunk in. There were, and are, many in the West who believed that expansion would store up resentment for the years ahead, but demonstrations of bitterness, while possibly satisfying for Russia, were self-defeating. Russia would inevitably have a large place in the future security of Europe, but it would have to forge a cooperative relationship with an alliance which had survived the end of the Cold War, as the Warsaw Pact had not. What was not certain, however, was whether Russia's leaders would see that and act appropriately.

The broadly shared opposition to NATO enlargement produced the most intense public debate over the direction of Russia's foreign policy. This time it was backed and led by a strong and very popular figure across the entire Russian political spectrum – Foreign Minister Yevgeny Primakov. This had a unifying effect on Russian policies towards NATO. Primakov personally took control, coordinating policy-making between various institutions and political parties.

Each political party, however, opposed enlargement on different grounds. Although the majority of pro-communist forces and the armed forces saw enlargement in purely military terms as a direct threat to Russia's security, other centrist and reformist politicians saw the threat from NATO enlargement mainly in economic terms, as delaying Russia's integration into the world market. For Russia's political élites, the shift in Western attention from support for Russian reforms towards support for Central and East European countries symbolised the West's long-term commitment to these countries, leaving Russia unstable and insecure for years to come. Thus, despite strong anti-Western rhetoric from the Russian leadership over NATO enlargement, its opposition did not overshadow continuing relations with international financial institutions and bilateral economic cooperation with Western countries.

The effect of NATO enlargement is one of the many unknowns in Russia's future. The political class had an enormously difficult task in 1996: to mediate the shock of continuing economic decline, of a rapid drop in status and the dissolution of previous power relationships. In many ways it fared badly, especially as President Yeltsin's incapacity put many developments on hold. But Russia moved towards the twenty-first century with no major catastrophe, apart from the Chechen war. It has learned some lessons, and has not foreclosed on either democracy or economic reform. As 1997 began, Russia remained a country in waiting.

Can Peace Last in Bosnia?

It once was that the coming of spring signalled an inevitable escalation of military confrontation among the warring factions in Bosnia-Herzegovina. The Bosnian war started in spring 1992; the Serbs consolidated their gains in spring 1993; the newly created Bosnian–Croat Federation attempted to reverse the military tide through limited counter-offensives in spring 1994; and the Serbs sought to end the war with a final, brutal attack on Muslim enclaves in eastern Bosnia in late spring 1995. It was the brutality of these later attacks – notably that around the UN 'safe area' of Srebrenica – that finally forced the international community, and particularly the United States, to seek to end the war. A forceful NATO bombing campaign in September 1995, combined with a ground offensive against the Bosnian Serbs by the newly trained Croatian and Bosnian government armies, produced a peace agreement negotiated in Dayton, Ohio, in November 1995. The 60,000-strong NATO-led IFOR troops that deployed in December 1995 to enforce the General Framework Agreement for Peace in Bosnia and Herzegovina (the Dayton Accords) ensured that spring 1996 was the first in five years with no major military activity by the parties concerned.

There is unlikely to be a return to war in spring 1997 – the cease-fire has lasted for over 18 months, much-needed economic reconstruction has begun to show initial results, the political leadership in Bosnia is beginning (however reluctantly) to address common problems, and over 30,000 troops remain as part of the NATO-led SFOR to ensure that a return to war remains a very costly option (see map, p. 238).

Yet, if war is unlikely to return to Bosnia in the immediate future, neither has peace been fully secured. Notwithstanding 18 months without shots being fired, a large international military presence, and an international community committed to peace in Bosnia, the prospects for the peace agreement – and for Bosnia – remain highly uncertain. Major aspects of the Dayton Accords remain unfulfilled; separatist and nationalist sentiment is undiminished and continues to tear at the fragile fabric of the Bosnian state; and turmoil in the neighbouring countries – especially in Serbia and Albania, but also to a lesser extent in Croatia – poses great challenges for peace and cooperation in Bosnia. By spring 1997, the forces of integration and peace, strongly supported by the international community, continued to vie for dominance with the forces of disintegration and conflict that persist both within Bosnia proper and within the countries that surround it.

A Mixed Record of Implementation

The Dayton Accords were based on two realities: first, all parties agreed that however unsatisfactory a negotiated peace in Bosnia, such a peace was preferable to continued war; second, the Serb desire for independence had to be accommodated within a framework that both preserved Bosnia's territorial and political integrity, and recognised the need for justice for the victims of what had been a particularly brutal war. The fundamental realities of the peace and the contradictions within it ensured that implementation of the Dayton Accords would not be easy and could possibly even flounder over the inevitable contradictions between military separation and political integration that the international community as well as the parties confronted.

These included the contradictions inherent in any country composed of two strong entities and a weak (barely existing) national government, a country with three armies but one foreign policy, two customs policies but one foreign-trade policy, two budgets but one central bank, and three presidents and rotating co-chairs of the Council of Ministers. From this perspective, the progress made by March 1997 in implementing the peace agreement was remarkable. Progress was most notable in the military sphere, but it was also seen in the economic, political and humanitarian spheres – albeit with descending degrees of success.

Military Implementation

With few exceptions, the detailed provisions and timetable for implementing the military aspects of the Dayton Accords outlined in Annex-1A of the agreement have been carried out in full and on time. IFOR began deployment to Bosnia on 16 December 1995 to enforce compliance with Annex-1A. It also supervised and, where necessary, enforced the cease-fire, the transfer of territory between the entities, the separation of forces and creation of a zone of separation (ZoS) along the Inter-Entity Boundary Line (IEBL), the marking of minefields in and around the ZoS, the demobilisation and confinement to barracks of combatants, and the cantonment of heavy weapons.

Figure 1 *Principal Obligations Under Annex 1-A of the Dayton Accords*

- Monitor cessation of hostilities along agreed cease-fire lines
- Withdraw foreign forces including 'individual advisers, freedom fighters, trainers, volunteers and personnel from neighbouring and other states'
- Redeploy forces of Bosnia and Herzegovina and the Entities into Zones of Separation extending 'for approximately two kilometres on either side of the agreed cease-fire line'
- Withdraw heavy weapons and forces to cantonment/barrack areas or other locations designated by the IFOR Commander within 120 days of the transfer of authority from the UNPROFOR to IFOR
- Establish Joint Military Commission to serve as the central body for all Parties to the military annex
- Provide for prisoner exchanges

Military implementation succeeded with no significant incident and the guns have been silent since October 1995. Compared to what preceded it, this was a remarkable and noteworthy success. Once achieved, it freed personnel and resources to support a wide variety of civilian tasks – including reconstruction, supervising national elections and protecting public security. And while NATO judged that stability could not be assured without an outside military presence, it decided that a follow-on capability to IFOR could manage with half IFOR's peak strength.

A second aspect of military implementation involved stabilising the military balance in Bosnia, although neither IFOR nor SFOR assumed any responsibility for securing this objective. Annex-1B of the Dayton Accords provided for the lifting of the arms embargo against Bosnia and the other former Yugoslav states, in place since September 1991, and mandated negotiation of three different arms-control agreements – including confidence- and security-building measures to be negotiated between the Bosnian parties (Article II); a CFE-type agreement reducing heavy weapons to be negotiated between the Bosnian parties, Serbia and Croatia (Article IV); and a regional arms-control agreement to be negotiated among countries in the wider region (Article V). Details and a strict timetable for negotiating the first two accords were agreed at Dayton and were to be negotiated under OSCE auspices. The Article II agreement was finalised in February 1996 and its execution has been largely satisfactory. The Article IV agreement proved more controversial to negotiate, but was finally signed on 14 June 1996. This Agreement on Sub-Regional Arms Control reduces Serbian tanks, armoured combat vehicles, artillery, combat helicopters and aircraft to about 75% of their mid-1996 totals. Bosnia and Croatia were each allowed to maintain about 40% of the level of Serb holdings in these categories. Two-thirds of the Bosnian post-reduction holdings were allocated to the Federation (Bosnian and Croat) forces, and one-third to Bosnian Serb forces. Meeting the agreement's requirements will demand significant reductions in Serbian holdings (well over 1,000 treaty-limited equipment – TLE), huge reductions in Bosnian Serb forces (over 2,000 TLE – about 70% of its wartime inventory), virtually no reduction in Croat forces, and large reductions in Federation artillery coupled with the possibility of increasing other TLE (notably tanks and ACVs) quite substantially.

The disparity in TLE holdings and reductions mandated by the accord has complicated its implementation. Although Serbia has agreed to reduce its forces even on an accelerated timetable (in part because of a US promise to assist financially), the Bosnian Serbs have sought to circumvent the agreement by abusing exemptions and substantially under-reporting holdings, especially of artillery. At the same time, the US-led train-and-equip programme for the Federation forces, while successful in eliminating Bosnian dependence on radical Islamic and Iranian weapons supplies, has complicated efforts to enforce Bosnian Serb compliance. This $400m project provided the Federation with some 75 tanks, 80 armoured combat vehicles, 78 long-range artillery pieces, 45,000 M-16 rifles, sophisticated communications equipment, and extensive training by retired US army officers. Nevertheless, arms-control progress was made when the Bosnian Serbs agreed in early 1997 to destroy some 1,000 TLE in the first reduction phase in return for US financial assistance with the overall process.

Economic Reconstruction

Bosnia emerged from the war a devastated country – in early 1996 its gross national product (GNP) was estimated at 10% its pre-war size, per capita income was 25% of its 1991 level, over 70% of the industrial plant had been demolished, and unemployment was 80% and rising as soldiers demobilised. On top of the war damage, Bosnia in 1996 faced the need to move from a centrally planned economy to one based on free enterprise, private ownership and the market. And if that did not pose a sufficient challenge, the political split in Bosnia was mirrored economically: three currencies circulated simultaneously; two (if not three) budgets existed to account for public services; and foreign – primarily Serbian and Croat – direction of parts of the Bosnian economy further weakened central control.

Given this background, Bosnia's economic recovery was remarkable. Although slow to start in the first half of 1996, reconstruction was fully under way by early 1997. The progress was aided by an infusion of substantial portions of the $1.8bn pledged by the international community for reconstruction in 1996 and by improved coordination among donors through the establishment of an Economic Task Force and the appointment of a senior Deputy High Representative for Economic Reconstruction. Moreover, in part to facilitate its own movement, IFOR greatly assisted the reconstruction effort – notably by building roads, repairing bridges, reconstituting the rail network and contributing to the early opening of Sarajevo airport in August 1996. In the Federation, industrial output in the first half of 1996 was 70% greater than a year earlier and net wages quadrupled from DM40 to DM160 a month. Unemployment fell to 65% according to official statistics, although an opinion poll conducted in late 1996 suggested that almost 70% of adults were in full-time employment.

Significant challenges remain mainly because the political foundation for economic recovery is virtually absent. Three in particular require attention:

- Privatisation and regularising property rights requires agreement on a legal structure within and between the entities. Without such a structure, the black market, corruption and criminality – already substantial – will supplant the official economy.
- Economic recovery requires an infrastructure that enables goods and services to move – within entities, between them and across international borders. Although devastated transport and communications networks pose a problem, political obstacles prevent establishing economically rational networks – such as the ability to move freely between entities, to telephone and fax directly between Pale and Sarajevo rather than through Belgrade or by satellite, and to charge uniform custom fees and duties for goods entering the country.
- Easy access to capital would provide the single greatest engine of economic growth. Unfortunately, Bosnia still lacks agreement on a central bank and financial regulations that make capital economically available rather than politically determined or proscribed. Neither foreign trade nor inter-entity commerce can flourish without a common currency and central fiscal policy. The absence of either will stymie any potential economic recovery.

Political Implementation

The weakness of joint political institutions poses a major problem for Bosnia's recovery – indeed, for its survival. While the letter of the Dayton Accords may largely have been followed – with national elections held in September 1996, and joint institutions established – the political reality is very different from what a literal reading of the Accords would suggest. Predictably, elections held so soon after the war had ended produced nationalist victories within all three ethnic communities. While fraud and other irregularities marred the final results, they were not significant enough to alter the outcome – which was produced by a high voter turnout, crucial logistical support by IFOR, and quiet cooperation by all parties to ensure an almost complete absence of violence. Nationalist leaders were elected to the three-person Bosnian presidency, and ruling nationalist parties won major gains in the Bosnian and entity parliamentary elections. Municipal elections originally scheduled for the same time were postponed when it became clear that the Bosnian Serbs and Serbia had exploited registration provisions in an attempt to secure by vote the Bosnian Serb presence that ethnic cleansing had established in former Muslim villages and towns (including Srebrenica and Brcko).

Although the September 1996 elections succeeded in establishing the constitutional structures agreed to at Dayton, these have so far failed to operate effectively. The Presidency and Council of Ministers meet periodically, but hardly ever to govern in any real sense of the word. The location of meetings, ethnic distribution of seats and rights, and the symbols of Bosnian unity (flag, seal, oath of office) long remained the primary topics of discussion. In the end, no constitutional structure can by itself reconcile the Serb desire for independence with the Bosnian insistence on unity and, indeed, these structures appear to have failed in resolving such apparently irreconcilable differences. This suggests that cooperation at the top is unlikely so long as nationalist leaders occupy these positions; instead, cooperation may have to be forged from below – through the day-to-day interaction and integration that a recovering economy requires.

Although economic integration and recovery is likely to give most people a stake in further pursuing peace, progress at the grass-roots level will require local political reform. Many towns and municipalities remain in the hands of local thugs, corrupt black marketeers, and others who benefited financially from the war. To take the most extreme example, western Mostar is run by the Bosnian Croat mafia, which is accountable to no one. Municipal elections were intended to give people the means to oust the thugs from power. But having been postponed once following Serb abuse of the registration process, the date has since changed three times because the OSCE lacks the administrative wherewithal to ensure that elections can be relatively free and fair.

Moreover, the international community's excessive focus on the election date and the logistical support necessary for conducting elections (a focus spurred in part by NATO, which would like elections to be held soon to allow a further drawdown in troops), continues to ignore the more fundamental problem. Conditions for municipal elections simply remain far from ideal. Immature opposition parties, government-controlled media, and a wide disparity in resources mean that even elections conducted without incident will be unlikely to bring about the democratic changes hoped for. The international community would do well to focus on laying the minimal necessary conditions for democratic elections rather than concentrating on their logistic and administrative details.

Humanitarian Considerations

The particular brutality of the Bosnian war meant that implementing the humanitarian provisions of the Dayton Accords – those dealing with refugees, minority rights and war crimes – would be at once the most urgent and yet the most difficult to enforce in a short period of time. Critics of Dayton and of the international community's efforts in Bosnia are generally right in pinpointing the key shortcomings:

- only some 220,000 out of 2.4m refugees and displaced persons have returned to Bosnia since the Dayton Accords were signed, and a further 100,000 (mostly Serb residents of Sarajevo) have since become displaced;
- even limited attempts to resettle abandoned villages in and around the zone of separation have been thwarted;
- the wilful violation of minority rights continues, notably the persistence of ethnic cleansing through the forced expulsion of minorities from their homes (a practice occurring in all three ethnic communities) and the deliberate obstruction of people's right to move and settle freely through campaigns of terror and destruction of housing; and
- the failure to arrest indicted war criminals (only seven of whom are in custody in the Hague to stand trial before the International Criminal Tribunal for the former Yugoslavia – ICTY), even though many of those indicted live openly in Bosnia, Croatia and Serbia and some continue to exercise power either openly or barely hidden behind the scenes.

At the same time, these are core issues – where people live and how they were treated was what much of the war in Bosnia was about. The resolution of core issues is never easy and certainly not in the initial stages of implementation of an ambitious – and ambiguous – peace agreement. In 1997, the key challenge facing peace in Bosnia is to ensure that these issues are addressed. The UN High Commission for Refugees (UNHCR) planned to make it easier for 200,000 displaced persons to return to majority areas.

The UNHCR and other international organisations – including SFOR, the police task force and the High Representative's office – have been working with the parties to agree on a process of gradual minority returns, especially in economically viable areas of the ZoS. The major powers were also considering ways in which to support the ICTY more effectively in 1997 – including ways to ensure the arrests of indicted war criminals. While SFOR would not be involved in a deliberate effort to hunt down indictees, the large military presence in Bosnia may give separate law-enforcement agents the degree of security they need to apprehend at least some of the indicted.

Key Challenges Remain

The absence of strong, capable and legitimate governing authorities at the local, entity and national levels creates a slew of problems that will be difficult to address. In part, this absence – symbolised by the chequered Dayton implementation record – reflects real differences over the nature of the Bosnian state, including the role of its two entities; however, the risk that the current Bosnian political authorities will disintegrate is also due to petty and incompetent leadership of the three ethnic communities. In 1997, the Bosnian state – including the two entities – risks becoming increasingly dysfunctional and may ultimately implode because of growing differences between the parties on the nature, purpose and scope of central-government action, increasing disagreement among the leaders of the two entities, and unbridled crime and corruption at all levels of government.

Although the joint institutions – a three-member Presidency which meets twice a week and a Council of Ministers which convenes regularly – were established only months after the September agreement, these bodies have failed to govern in any real sense of the word. Serb and Bosnian disagreement on the competencies of the national government – whether over fiscal policy, foreign policy or trade policy, which nominally fall under central-government control – have blocked action, effective or otherwise. Instead, meetings focus on symbolic issues that underscore the tension between the separatist and integrationist desires of the government's members. Moreover, any decision, however small or insignificant, is first referred back to the entity government for informal vetting and approval – assuring a slow, tedious, ineffective and largely insignificant governing process.

The Brcko area, situated in the Posavina corridor – a 40km strip linking Serb-controlled areas in western and eastern Bosnia – nearly derailed the Dayton Accords in November 1995. For the Muslim–Croat Federation, Brcko represents an important economic, political and strategic prize whose possession would undermine the political cohesion of the Serb Republic. The Dayton Accords postponed a decision on the fate of this most contentious and potentially explosive issue until arbitration could take place. Annex 2, Article V, stipulated that arbitration would be binding and a decision reached within one year of signing the peace agreement – in other words, by 14 December 1996. This decision was, however, postponed until February 1997 by the presiding arbitrator, Roberts Owen, appointed by the International Court of Justice. At a meeting of the arbitration panel in Rome on 15 February 1997, Owen announced that a final decision on the strategic town would be further delayed until 15 March 1998, and that in the interim an 'international supervisor' would be appointed to assist economic reconstruction and the resettlement of refugees. A US diplomat, Robert Farrand, was appointed to this post by the High Representative Carl Bildt on 7 March 1997. These two postponements over a decision on Brcko's future illustrate the problem of solving what appears to be an irreconcilable problem; further delay is unlikely to make it any easier.

A Disintegrating Republika Srpska?

The weakness of the central state is compounded by the weakness – even near-collapse – of the entity authorities. Most immediately worrisome is the fact that, despite its strong desire for independence, the Republika Srpska (RS) is on the verge of collapse. It is economically destitute, increasingly weak militarily, and riven by petty political infighting among a small group of leaders.

The Republika has enjoyed few of the economic fruits that Bosnia's limited recovery has produced. While industrial output and wages rose and unemployment fell in the Federation in 1996, economic statistics in the RS moved in the opposite direction. Its largely agricultural and primary-resource economy remains devastated – with land-mines preventing many fields from being worked and with industrial mines still essentially closed. Above all, however, the RS suffers from lack of access to capital. Because it refuses to support central institutions and continues to create numerous other compliance problems, Sprska has received just 2% of the aid expended in Bosnia by the international community since the Dayton Accords were signed. Moreover, while Srpska relied on Belgrade for economic and financial support throughout the war, the collapse of the Serbian economy has closed off this source of revenue as well. Unless capital and assistance begin to flow to Srpska soon, its economic collapse is almost certain.

Militarily, the RS is little better off. The Yugoslav National Army (JNA) continues to be responsible for paying officers and soldiers, but is itself financially squeezed and can no longer maintain even a much smaller RS army. In late 1996, moreover, the newly elected RS political elite successfully challenged the old army leadership for control over the armed forces. As a result, those commanders who had most successfully prosecuted the war – including the commander of the RS Army, General Ratko Mladic – were relieved from duty and replaced by more politically pliant (although not necessarily more militarily capable) commanders. The internal military reshuffling, combined with the lack of resources, has left a demoralised military in possession of ageing weapons. There are believed to be no RS military units currently capable of sustained and coordinated military action, with the possible exception of defending the immediate area in which the unit is deployed.

In addition, Srpska continues to be riven by petty political infighting. The leadership in Pale remains very much beholden to former 'President' Radovan Karadzic. Although he was nominally removed from public office and influence as a result of negotiations between US Assistant Secretary of State Richard Holbrooke and Serbian President Slobodan Milosevic and Bosnian Serb leaders in July 1996, Karadzic continues to exercise political power behind the scenes – in particular concerning the possibility of devolving power from the RS to the national government. A more moderate leadership located in Banja Luka – sometimes represented by the elected President of Srpska, Madame Biljana Plavsic – has yet to succeed in marginalising the Pale power clique around Karadzic. Such manoeuvring for power and influence among the 50–100 Bosnian Serb officials with any influence has deprived Srpska of a leadership devoted to resolving the deep-seated problems confronting the entity. Without such leadership, further disintegration is likely.

Can the Federation Muddle Through?

Although it suffers from less severe problems than Srpska, the Bosnian–Croat Federation is hardly a model of effective and cooperative governance. The Federation is still a body that exists more on paper than in reality. Although there is a nominal Federation government, effective action is still largely funnelled through separate Bosnian and Croat channels. What cooperation does exist is enforced by the international community. Economic assistance is predicated on central Federation institutions and cooperation between the two communities, and the military aid provided under the US train-and-equip programme is conditional on defence integration. In addition, a high-level forum co-chaired by the United States and the High Representative's office continues to encourage the two sides to share power and cooperate in common endeavours. There has been some success – notably in establishing a single customs policy and, after eight tries, the formal dissolution of the Croat-run Hercog-Bosna 'republic' – but hardly sufficient success to ensure a fully functioning entity government.

The Federation's problems are compounded by an increasingly dissatisfied Bosnian leadership. It was clear at Dayton that the Bosnians were the least satisfied party – and little since the agreement was initialled has convinced the leadership that a just peace is within reach. Lack of progress on those humanitarian issues judged fundamental to the Bosnian leadership because they provide some measure of restitution for the main victims of the war remains a constant source of dissatisfaction with Dayton. The result has been coordinated attempts to impel progress on these issues – notably through the forced return of refugees to strategic areas in the ZoS. The violence that predictably followed in such hamlets as Jusici and Gajevi, in other areas in the Posavina and Zvornik corridors and in the Sapna Thumb region has made the return of refugees to minority areas even less likely, further feeding Bosnian dissatisfaction.

The combination of the continued dysfunction of the central governing structures, the potential collapse of Republika Srpska, the weakness of the Federation, and growing dissatisfaction with the status quo on the part of the Bosnian leadership poses a grave challenge to the long-term viability of the Bosnia envisaged at Dayton. While the international community remains fully engaged – chivvying all sides to cooperate – it is unlikely that these internal fissures will erupt violently. However, unless these problems are addressed and resolved, the international community will have to choose between staying militarily engaged for years, or even decades, and leaving the Bosnians to their own devices, even at the risk of violence erupting once again.

The Middle East

The intransigence of the ideologically driven *Likud* government in Tel Aviv and Palestinian Chairman Yasser Arafat's weakness combined during 1996 and early 1997 to cast the once-promising Arab–Israeli peace process into doubt. Arafat has always had the more difficult task. His ability to control extreme Arab parties, intolerant of the idea of peace with a continuing Israeli state, has depended on showing that advances in the peace agreement would result in a better economic and political life for the Palestinians. This hope has been stymied by Israeli Prime Minister Binyamin Netanyahu whose own preference for what he believes is firm security over a dicey peace has led him to delay, and try to change, the carefully wrought Oslo Accords which offered a way forward.

The result was a return to rock-throwing street demonstrations against Israeli soldiers and suicide bombings against Israeli civilians in 1996 and early 1997. As the violence began to mount, the Israelis retaliated with tear gas, water cannon, rubber bullets and then live fire. The deadly cycle of 'tit-for-tat' retaliations was once again under way. Although outside powers, particularly the United States, tried to play a constructive role, this will be unachievable without strong leadership from both Israelis and Palestinians. Since this seems unlikely, hopes for an early meeting of minds between the two sides have receded further.

The US exhibited strong leadership over Iraq during 1996. Saddam Hussein's move into the protected Kurdish territory in northern Iraq was met by US bombing raids and an adjustment to the southern no-fly zone. France, however, showed its disagreement with present US policy in the region by refusing to monitor the extended zone. Some Arab countries, too, have begun to express their doubts about the need to maintain economic pressure against Saddam Hussein with quite the vigour exhibited up to now. These developments have given Saddam Hussein greater confidence that he can drive a wedge between his international opponents to his own advantage. But there are signs that his grip on internal control may be weakening; if it slips much more there could be a change of government – for better or worse – within Iraq in the not-too-distant future.

The Peace Process at a Crossroads

No one thought it would be easy, but developments in 1996 have highlighted the many difficulties that must be overcome before there is any hope of resolving the Arab–Israeli conflict. At one time, progress towards resolving this conflict seemed unstoppable, merely a matter of time before all the major issues would be settled. After all, the regional actors lacked any realistic or affordable military alternatives for dealing with their dispute. More than once in the last year, however, negotiations have stalled and threatened to break down, leaving the area on the verge of yet another wave of extremism and violence.

While what has been achieved between Israel, the Palestinian Authority (PA), Jordan and Egypt did not completely unravel in 1996, the momentum necessary for a comprehensive deal and normalisation of relations has indeed weakened. This has occurred just as the parties are about to confront the most difficult core issues: the future of Jerusalem; the status of Palestinian refugees; Israeli settlements; security arrangements; final borders; and water resources. An outline for dealing with these issues has scarcely been proposed, let alone agreed. The emergence of a new Middle East devoid of conflicts and wars will take time, and more determined diplomatic efforts by the parties themselves, as well as by outside actors, will be needed to prevent the situation deteriorating further. Without such efforts to rescue the peace process from the stalemate it had reached by March 1997, there will inevitably be a return to confrontation.

Netanyahu at the Helm

The May 1996 national elections were a watershed in Israeli politics and foreign policy. The elections were held against the background of an intensely divided population, epitomised by the assassination in November 1995 of then Labour Prime Minister Yitzhak Rabin by an extremist Jewish student. And, for the first time, the voting public was to elect the new Prime Minister directly.

Foreign policy in general, and peace-process-related issues in particular, loomed large during the electoral campaign. Shimon Peres, succeeding Rabin as leader of the Labour Party, promised to continue his predecessor's policies, while Binyamin Netanyahu, leader of the *Likud*, agreed to continue negotiations with the Palestinian Authority provided the PA acted decisively to control militant groups. More important was Netanyahu's emphasis on personal and national security over every other consideration.

For at least two months after Rabin's assassination, Peres enjoyed a lead of 10–20% over the opposition parties. This changed dramatically, however, when a wave of suicide bombings by militant Islamists swept Israel in late February and early March 1996. This outbreak of violence, and the government's powerlessness to prevent the high death toll among civilians, increased doubts in Israeli society about the government's ability to protect its citizens. The *Likud*'s campaign played on these fears and the gap between Peres and Netanyahu was soon eliminated. Despite last-minute attempts by Peres to demonstrate his tough approach to security (for example, by mounting the perhaps misguided *Operation Grapes of Wrath* in April 1996 against *Hizbollah* in Lebanon), he was unable to halt the erosion of his credibility among Israeli Jews and, at the same time, alienated many Israeli Arabs.

Binyamin Netanyahu emerged victorious over Shimon Peres by the astonishingly narrow margin of about 30,000 votes – less than 1% of the overall total turnout – but with an 11% lead among Jewish voters. As he moved from leading the opposition to heading the government, there was much debate on what peace policy Israel would follow under his leadership. Would Netanyahu prove to be an ideologue or a pragmatist?

Some observers expected Israel's new Prime Minister to pursue a hardline policy towards the Arabs, faithfully reflecting the revisionist ideology of his right-wing coalition. This ideology stresses the primacy of national security and the need to maintain most of the West Bank under Israel's sole control for religious and strategic reasons. Even if Netanyahu himself wished to make some fundamental changes in *Likud*'s foreign-policy orientation by granting territorial or political concessions to the Palestinians and Syrians, he could not completely ignore the right-wing hardliners within his party. Nor could he ignore the leaders of the small, but vital, religious parties whose political backing would be essential to keep the coalition in power.

Others argued that Netanyahu came to power promising to pursue peace with Israel's Arab neighbours without compromising Israeli security. Regardless of ideology, he would have to consider the divisions in Israeli society towards the terms of the peace. In short, he had to be aware that almost half the Israeli electorate opposed his policy. Maintaining Israel's high economic growth rate would clearly require *Likud* to moderate its policy towards the peace process. This would become politically feasible if Netanyahu brought the opposition Labour Party into a grand coalition, or if he made some significant concessions in negotiations with the Arabs which Labour, even though in opposition, would be expected to back.

Netanyahu's election as Prime Minister aroused political fears in the Arab world where many hoped Peres and his formula of 'land-for-peace' would win. To decide a strategy to deal with Netanyahu's 'peace-for-peace', Arab leaders, with the exception of the uninvited Iraqi President Saddam Hussein, held an emergency summit in Cairo on 21–23 June 1996. They affirmed their commitment to peace 'as a strategic option', and called for a resumption of the peace talks on all tracks. They also insisted on Israel's withdrawal from all the Arab occupied territories and the creation of an independent Palestinian state with East Jerusalem as its capital. Moreover, they threatened to reconsider steps already taken to normalise relations with Israel should the new government renege on the land-for-peace formula. In taking these positions, Arab leaders tried to avoid creating more divisions in their ranks while affirming the need to see how the new Israeli government intended to act before taking more decisive diplomatic action. But the willingness of Arab leaders to judge Netanyahu on his actions, rather than by his early rhetoric, was not soon repaid by any early attempt by the *Likud* government to calm the Arabs' worst fears about the fate of the peace process.

Indeed, by the end of September 1996, tensions were rising fast. The redeployment of Israeli forces from Hebron, supposed to have begun in March 1996, was still not in sight. Additional strains produced by Israel's closure of the Palestinian territories in the West Bank and Gaza highlighted the dangers inherent in a stagnating peace process. At this juncture, with the atmosphere becoming increasingly heated, the Netanyahu government took the controversial decision to open a new entrance to an existing tunnel under the holy sites of the old city of Jerusalem. Palestinian rage exploded in street riots. More alarming, however, was a fierce exchange of fire on 26–27 September between the Israeli forces and Palestinian police which led to the death of over 65 Palestinians and 15 Israelis. Palestinians viewed the crisis as an example of Israel's high-handed unilateralism and insensitivity, while Israelis were alarmed by what they saw as the PA's sanction of its police force turning Palestinian guns against them. The riots demonstrated the strong potential for violence that exists so long as agreement is not reached and consolidated.

In an attempt to diffuse the Palestinian–Israeli tensions, to prevent wider repercussions in the region and to restore momentum to the peace process, US President Bill Clinton called a Middle East summit in Washington in early October with Israeli, Palestinian and Jordanian leaders. Egypt's President Hosni Mubarak chose not to participate since he did not expect much from such a hastily called summit. His doubts proved to be justified as the talks produced few concrete results. Israel insisted on keeping the Jerusalem tunnel open and the political stalemate in its talks with the PA persisted. Nevertheless, the summit helped calm the immediate situation and provided a breathing space within which to increase external and internal pressures on Netanyahu. As fears grew that the situation would deteriorate further, the majority of Israelis who had welcomed the peace already won (polls consistently put this figure at about 70% of the population) made it clear that they did not wish to see a return to confrontation and conflict.

The Hebron Accord

Since the signing of the Oslo Accord in September 1993, one of Israel's principal demands has been that the Palestinian Authority curb as much as possible, if not eliminate, terrorism against Israeli targets. During 1996, the PA had recognised the importance of acting to alleviate Israeli concerns and it successfully detained hundreds of suspected Islamic activists or sympathisers. The Israeli security services recognised this and impressed on

Netanyahu the importance of accepting Palestinian Chairman Yasser Arafat more seriously as a partner in negotiations. They particularly stressed the need to redeploy Israeli troops from most of Hebron immediately. Failing to do so, they argued, particularly given the economic burden on the Palestinians caused by Israel's closure of the territories, would undermine the PA's ability to maintain order. And there was no likely alternative leadership in the territory that could prove more advantageous to Israel.

The Labour government had reached an agreement in September 1995 with Arafat on redeployment from Hebron, but its implementation had been postponed, first because of the spate of suicide bombings in early 1996 and then because of the Israeli elections in May. During the electoral campaign, Netanyahu insisted on revising the terms of the Labour-negotiated accord before committing himself to redeployment. Renegotiating the terms with the PA, however, was extremely difficult; the negotiations lasted for approximately four months, during which they teetered repeatedly on the verge of collapse.

Figure 1 *Main Aspects of the Hebron Accord*

- Israeli forces will redeploy from most of Hebron within ten days of signing the Accords
- The Palestinian Authority will assume responsibility in the areas from which Israeli forces have withdrawn
- Israel will remain responsible for the security of Jewish settlers at the City of Hebron and the Tomb of the Patriarchs
- Israel and the PA will establish joint intervention teams and joint patrols to function in the Heights overlooking a designated area where the Jews of Hebron are living
- Specific restrictions apply to the number of Palestinian police and the personal arms they are allowed to carry
- Israel will agree to open two main streets closed earlier for security reasons although they pass through the Israeli-controlled area

During these discussions, Netanyahu had to consider external actors, primarily the United States. As Israel's main international ally, Washington's views and perceptions could not simply be ignored by Netanyahu or, for that matter, by any Israeli Prime Minister. Although the US has consistently refused to compel Israel to withdraw from the occupied territories, Washington had a stake in preventing the Arab–Israeli peace process from collapsing. To avoid such an outcome, US envoy Dennis Ross formulated compromise proposals during the lengthy talks leading to the signing of the

Hebron Accords on 15 January 1997, and thus succeeded in forcing an initially reluctant Netanyahu to close the Hebron deal on terms not far from those previously proposed by Labour.

Despite his efforts to cater to right-wing views on this issue, Netanyahu's ultimate acceptance of the Hebron Accord infuriated the religious right in Israel to whom giving Arafat control over most of the city was an act of betrayal. Although he won the support of his cabinet by 11 votes to seven, the cost of the deal was indicated by the resignation from the cabinet of Benjamin Begin, son of former Prime Minister Menachem Begin, in protest. Another former Prime Minister, Yitzak Shamir, accused Netanyahu of betraying *Likud*.

Just when Netanyahu risked losing right-wing support over Hebron, he needed their backing more than ever. He was accused of appointing Roni Bar-on as Attorney General on 10 January 1997, allegedly to gain the support of the religious *Shas* party for the Hebron Accord. In return, *Shas* leader Aryeh Der'i, who was under criminal investigation, would be treated leniently by the pliable Bar-on. In the event, Bar-on, whose credentials for such high office were dubious, lasted less than two days in the post. But the threat of a judicial investigation left Netanyahu politically vulnerable.

In an attempt to narrow the gap between himself and his rightist critics, Netanyahu engaged in a balancing act by expanding settlements in Jerusalem and securing Israel's control over its outer suburbs. Israel's decision on 26 February 1997 to build 6,500 new Jewish settlements at Har Homa (or Jabal Abu Ghneim), a hill top between East Jerusalem and Bethlehem, was part of this strategy. This project is expected to increase the number of Jewish settlers by about 32,000 and to close the last corridor between East Jerusalem and the southern part of the West Bank. Netanyahu's decision paid off in a narrow political sense. The Land of Israel Front, which had denounced the Hebron Accord, hailed Netanyahu's 'commendable courage'. The cost was high, however. The decision caused uproar among the Palestinians and Arabs who saw it as another Israeli attempt to create irreversible facts on the ground before negotiations on the final status of Jerusalem and the West Bank, scheduled for late March 1997, begin. Netanyahu's 'compensation' to the Arabs – an offer to build homes for Arabs as well and the net transfer of 2% more West Bank land to Palestinian control – was derided by the Palestinians (see maps, pp. 244–45).

Arafat's Dilemma

As a rule, although the more extreme Palestinian groups have always opposed the Oslo Accords, they have been unable to overturn them through suicide bombings and political denunciation. Yet Arafat still faces serious challenges in his dealings with Israel. Israel continues to have the power to squeeze the enclaves under Arafat's control should he adopt policies it considers unacceptable. The Israeli settlement policies and the worsening conditions of Palestinians in these enclaves weaken the credibility of Arafat's repeated promises to his people of an independent Palestinian state with Jerusalem as its capital.

As the gap between promise and reality widens and the inability of the PA to force Israel to change its positions becomes clearer, Arafat has vacillated between massively suppressing the opposition and initiating a dialogue with it. Hundreds of people suspected of belonging to opposition groups were detained in PA prisons without charge or trial. But when this failed to move the negotiations along and tensions with Israel rose, Arafat took another route and invited ten Islamist and secular opposition groups for talks to reach a unified national programme as a way of exerting pressure on Israel. On 27 February 1997, he organised the Comprehensive Palestinian Dialogue Conference in Nablus where he met with representatives of his Palestinian rivals: *Hamas*, the Popular Front for the Liberation of Palestine and the Democratic Front for the Liberation of Palestine. He also released many of the imprisoned Islamist and leftist cadres.

Arafat himself made clear that the PA has no choice but to continue with the peace process, a stand which severely limits the likelihood of forming a united Palestinian front. Yet, Arafat also tried to show his radical critics that they had no alternative but to deal with him. At the same time, he has tried to maintain active links with any group, anywhere, that he hopes would create more pressure in favour of the peace process. He has entreated the Israeli peace camp and business community, which he hopes might force Netanyahu to reconsider some of his policies. Arafat also sought the assistance of Egypt and Jordan in December 1996 specifically to influence Israel's position on Hebron. The Palestinians as a whole have encouraged active European diplomacy to persuade the *Likud* government to moderate its policies and to demonstrate to the US that they can recruit strong support within the West for a more 'balanced' approach to the Middle East situation. In mid-March 1997, in response to the threat of new settlements around Jerusalem, Arafat called a meeting in Gaza with US, European and Arab diplomats to attempt to salvage the peace process and prevent Israel creating irreversible facts on the ground.

Arafat's bargaining position remains weak and his freedom of political action limited. Although many Israelis have strong doubts about Netanyahu's policies, this is not the case regarding Jerusalem, on which opinion has been settled in Israel for years. The US continues to stress that direct talks between the Israelis and Palestinians must be the primary focus of the peace process , and thus is reluctant to lean too hard on Israel and force an outcome against its will. In March 1997 alone, the US twice vetoed a UN Security Council Resolution condemning Israel's plans to build on Har Homa, even though Washington had itself expressed strong reservations about the plan. As long as Washington continues to see 'message carrying' as the limit of its diplomatic ambitions in the area, the negotiating pace is unlikely to increase.

Lebanon: First or Last?

If the peace process between Israel and the PA stalled in 1996, that between Israel, Lebanon and Syria remained only rhetorical for all concerned. In April 1996, Israel launched *Operation Grapes of Wrath* in Lebanon, aimed at destroying the *Hizbollah* infrastructure in retaliation for guerrilla attacks against Israeli settlements and the pro-Israeli Southern Lebanese Army. Although *Grapes of Wrath* was destructive, it was another inconclusive Israeli military campaign in Lebanon. After Israel's artillery shells killed a large number of refugees at the UN outpost at Qana, international pressure mounted for an immediate cease-fire. A 'monitoring group', including France, Israel, Lebanon, Syria and the United States, was established to investigate cease-fire violations. Far from succeeding in its aim, the operation ended with *Hizbollah* suffering negligible losses to its fighting cadres, but gaining a valiant image for its capacity to stand up to Israel's mighty war machine.

The failure and costs of the operation led some in Israel to resurrect the concept of a 'Lebanon-first' option. Israel has always argued that it has no territorial ambition in Lebanon. Those who advocate the 'Lebanon-first' option believe that if Lebanon were free to decide for itself how to negotiate with Israel there would be strong support in Beirut for such a deal if the security of the settlements in the north could be secured and the Lebanese state could extend its military and political control over southern Lebanon (which *Hizbollah* uses as a launching pad for its operations against Israel and the Southern Lebanese Army). Establishing a reliable security regime would significantly reduce Israel's human losses and the material cost of the protracted confrontation along the Israeli–Lebanese border. It could also provide Israel and Syria with an arena in which to test each other's political intentions on a regional level in preparation for the next stage of their transition from war to peace. But there was the rub. Most who advocate 'Lebanon-first' know that to Syria Lebanon is a base from which Israel can be harassed into making concessions direct to Syria. Hence the cynical attempt of 'Lebanon-first' for some: if Syria refuses to cooperate on the option, the blame for the continued stalemate and for delaying Israel's withdrawal from Arab land could be placed firmly on President Hafez Al-Assad's shoulders.

Indeed, Syria, which has clearly been the dominant military and political force in Lebanon since its intervention there in 1976, did not find it in its national interest to cooperate with Israel on the 'Lebanon-first' option. Damascus insisted instead on a strict linkage between the Lebanese and Syrian fronts, and tightened its control over the Lebanese political establishment to ensure it would echo the Syrian position.

In essence, Syria's position has been that Israel should be made to suffer in southern Lebanon until an agreement is reached to withdraw Israeli forces simultaneously from Syrian and Lebanese territories. Delinking the two fronts would seriously weaken Syria's bargaining position and remove what it considers its political leverage in its effort to regain the Golan Heights from Israel. Linking the settlement on the two fronts as an indivisible package is essential if Assad is to prevent Syria's position *vis-à-vis* Israel becoming marginalised.

Walking Down a Frozen Track

Talks on the Syrian–Israeli track have been at a standstill for the past year. Before the talks were suspended in February 1996, the Syrian and Israeli Labour government negotiators had made headway on some aspects of the Israeli troop withdrawal, on the security regime to be established along the Syrian–Israeli borders, and on normalising relations between the two countries. But the Netanyahu coalition government refused to accept that any of these apparently only verbal agreements were valid. It remains committed to its political platform which stipulates that the principle of land for peace does not apply to the Golan Heights, which *Likud* views as vital for Israeli security (see map, p. 243).

Syria seems not to be interested in negotiating under such a formula. It argues that Egypt did not consider US economic aid a sufficient substitute for all of Sinai, and that therefore Syria should not be bought off its position on the Golan by promises of economic aid if it flexibly interprets the amount of the Golan it wants back. Damascus has rejected Netanyahu's proposal to resume the peace talks 'without preconditions' and insists that any resumption of talks must be based on the understandings accepted by Israel's Labour government in US-sponsored talks with Syria, particularly those held at Wye Plantation in 1995 and early 1996. Damascus does not want to start negotiating again from scratch particularly with an Israeli government that adopts a harder line towards Syria.

During 1996, tensions mounted between the two countries until fears were expressed of possible conflict between them. In August, movements of Syrian forces in Lebanon and near the Golan Heights were widely reported in Israel, yet none of these movements really indicated a more offensive posture on Syria's part. Syrian test-firing of *Scud*-C missiles at the end of a large-scale training exercise also attracted external attention. Threatening statements were exchanged in a war of words between Israel and Syria in the last few months of 1996. Syrian Chief of Staff General Hekmat Al-Shehabi reminded everyone that despite Damascus' strategic commitment to peace, a military option still exists to support Syrian policy objectives. In response, Israeli Defence Minister Yitzhak Mordechai threatened that if tensions were to develop into war, Syria could expect a massive Israeli attack.

Despite its strong statements, Syria is unlikely to start a military confrontation with Israel under current regional and international conditions. Syria could never go to war with Israel alone, and it could probably not form an Arab military coalition against Israel. The circumstances which made it possible for Syria to take the initiative in 1973 are no longer there. Egypt is out of the military equation and Saudi Arabia is not in a position to use oil as a political weapon, as it had done before. The Arab order which provided the cover for the earlier war with Israel is in disarray. Syria's then international ally and arms supplier, the Soviet Union, has disintegrated and Russia, burdened by its many troubles, is not in a position to assume the Soviet role in the Middle East. Syria's military faces major problems in acquiring spare parts for its equipment and in upgrading outdated weapons.

Moreover, Israel's military power far exceeds that of Syria. Even though Syria's rhetoric has often been strong, its behaviour has been prudent, carefully taking into account the relationship between desirable ends and available means. Thus, over the past two decades, the Golan Heights has remained the quietest Arab–Israeli front in terms of violence, and may continue to be so.

However, tensions between states preoccupied with their security dilemmas can escalate out of control without either side actually intending to initiate a war. Clashes between Israel and *Hizbollah* in Lebanon may lead to larger clashes with Syrian forces in the Beka'a Valley. Bearing in mind the present official Israeli position that excludes any withdrawal from the Golan, the possible goal behind a limited Syrian escalation of tensions with Israel might be to activate an international diplomatic search for a settlement more than to liberate the Golan Heights militarily. As long as the stalemate over the Golan continues, Syria could still take risks to change the diplomatic equation. In any case, so long as the Golan stays entirely in Israeli hands any peace with Syria is impossible. Continued stalemate between Syria and Israel entails certain risks.

If Syria believes that it has nothing to gain from the current peace process, it may do everything within its reach to frustrate that process. Syria can use its alliance with Iran and its relations with *Hizbollah* in Lebanon to increase the costs to Israel of maintaining its presence in southern Lebanon. It can place obstacles on the Palestinian–Israeli track through its close ties with rejectionist Palestinian groups, particularly as this track is progressing slowly. Syria can also be expected to escalate its campaigns in the Arab world to impose a freeze on normalising relations with Israel. While new US Secretary of State Madeleine Albright is not expected to visit Damascus as much as her predecessor, Warren Christopher, one challenge for US diplomacy will be to find an acceptable formula for resuming Syrian–Israeli negotiations.

Obstacles to Normalisation

The pace of normalising relations between the Arabs and the Israelis continues to reflect the political difficulties hampering the entire peace process. The Middle East situation illustrates the primacy of politics over economics when it comes to integration in a region where there is either no peace – as in the case of Syria and Israel – or where the peace is a cold one – as between Egypt and Israel. Arab countries which have established trade relations with Israel during the last two years – such as Morocco, Oman, Qatar and Tunisia – have reacted negatively to the stalemate in Israeli–Syrian negotiations and the recurrent stagnation of Israeli–Palestinian talks. They have thus tended to put normalising economic relations with Israel on hold.

Professional associations, mostly dominated by Islamists and Arab nationalists, in a number of Arab states (particularly Egypt and Jordan) continue to take the lead in opposing normalisation with Israel. In Jordan, they mounted a vigorous political campaign in October and November 1996 against cooperation between Jordan's private sector and Israeli businessmen, and caused an industrial exhibition to be cancelled. Israel was also barred from participating in film festivals in Egypt. Similar pressures exist, to a lesser extent, in Morocco and Tunisia. Reportedly, King Hassan of Morocco, despite his well-known moderate attitude towards Israel, has declined to hold talks with Netanyahu in light of the prevailing political mood in Moroccan society.

The major difficulties facing the peace process threatened the Middle East and North Africa Economic Conference in Cairo scheduled for 12–14 November 1996. Policy-makers in Egypt had doubts about holding the Conference with Israeli participation under such conditions. Some were clearly in favour of postponing the Conference for three months to pressure Israel to expedite its troop redeployment from Hebron. However, in what has been described as an 'Egypt-first' policy, President Mubarak concluded that holding the Conference as planned would be better for Egypt's international credibility and for encouraging foreign investments in the country. The theme of the Conference was no longer 'Middle Eastern Economic Integration', but 'Building for the Future: Creating an Investor Friendly Environment'. Private-sector companies urged governments to accelerate the peace process to create a climate more conducive for economic development before the next Middle East and North Africa Conference, due to take place in Qatar in November 1997. Even the Middle East Bank for Economic Cooperation and Development, which was agreed three years ago, has not yet started work, reflecting the political obstacles confronting institutions for regional economic growth and integration.

The overall picture leaves little room for optimism. The final negotiations between Israel and the Palestinians due to start in May 1996 have been twice delayed: first, as a result of mounting tensions over Israeli settlements near Jerusalem and militant violence against Israeli civilians in Tel Aviv; and second, because of the Israeli national elections. Violent confrontation is also increasing across the Lebanese–Israeli border. Given the diametrically opposed positions and lack of mutual confidence between the main regional parties, there are few prospects for an early start to Israeli–Syrian negotiations. The new Middle East that had beckoned from beyond the horizon is sinking further from view, while the old conflict-ridden Middle East refuses to give up. The threat of a return to another 'no war, no peace' period of instability, punctuated by street demonstrations and suicide bombings, is spreading its shadow once again over the region.

A Shaky Year in Iraq

The survival of Saddam Hussein and his regime for another year ensured that in 1996 Iraq continued to exhibit many of the features that have become so familiar to the outside world. Domestically, repression, clan rivalries, military conspiracies, purges and the problems of a fragmented opposition were prominent. Internationally, Iraq's relations with the United Nations and with most of its neighbours were characterised by the fluctuating bouts of aggression and conciliation that have marked Iraq's actions since the end of the 1991 Gulf War.

Saddam Hussein steered his usual course in 1996, facing up to continuing sanctions while also attempting to exploit the opportunities that UN Security Council Resolution (UNSCR) 986 (popularly known as 'oil for food') promised his regime. In northern Iraq, ongoing conflicts between rival Kurdish factions in the 'safe haven' gave the President an opportunity to boost his political standing, reminding his subjects of the still-formidable powers at his disposal. Furthermore, the mixed international reactions to the regime's first major military foray into the Kurdish areas since 1991 underlined the growing tendency of some states to accept Saddam Hussein as a political fact of life and to deal with him on his own terms. As a result, the regime seems increasingly confident in its ability to play parts of the outside world off against each other.

Internal Challenges to the Regime

To maintain the appearance of legitimacy, general elections were held in March 1996 for the Iraqi National Assembly. In highly controlled conditions, 689 candidates competed for 220 seats (30 seats for the Kurdish areas that are beyond effective Iraqi government control were not contested). All candidates were vetted by the security authorities and, although not technically required to be members of the Ba'ath Party, they were obliged to support its aims and, most importantly, to support the 'Necessary Leader', Saddam Hussein. Predictably, the elections produced an Assembly that Saddam Hussein could rely on to rubber stamp the decrees which form the basis of most Iraqi legislation.

Effective power continued to be based on the President's extended family and on the clans that have been the hallmark of Saddam Hussein's rule. In March 1996, repercussions continued from the murders in February of Saddam Hussein's two sons-in-law who had unwisely returned to Iraq after their public defection to Jordan in August 1995. As a result, members of the Al-Majid clan turned on each other. Some sought to rehabilitate themselves with their kinsman Saddam Hussein, while others suffered for being too closely associated with the murdered sons-in-law. In a culture of family feud and blood revenge, it was dangerous to leave too many loose ends untied. The President was able to eliminate those he believed had betrayed him, but he was also aware of the potential danger to his regime should the widening circles of vendetta cause the clans closest to him to destroy themselves. Other clans and other conspirators are always waiting to exploit rifts at the heart of the regime.

This troubled state of affairs may have led to a serious challenge later in the year. In July, a reported attempt on Saddam Hussein's life may have been part of a wider plot within the Iraqi Army's officer corps. The President survived, but the arrests and executions that followed rapidly in August targeted some officers from the most prominent military clans of Mosul and adjoining districts. Although an indispensable part of the officer corps, the Mosul clans have never been wholly trusted by Saddam Hussein who fears that they and their like represent the greatest threat to his power.

This smouldering suspicion was encouraged in summer 1996 by the establishment of a powerful radio transmitter to Arbil and other parts of Kurdistan by the opposition Iraqi National Accord, based in Amman. With the protection and patronage of the Jordanian and US governments, this group, formed in 1990, gave prominence to senior Iraqi military defectors and claimed to be cultivating close links with discontented groups within the Iraqi military. These claims were probably exaggerated. However, Baghdad's sensitivity on this issue was demonstrated in August when, against the background of the reported arrests and executions of Iraqi officers and the increasingly public activities of the National Accord, a number of Iraqi diplomats were expelled from Jordan. They were accused of having been instrumental in an Iraqi plan to instigate riots in Amman in retaliation for the assistance Jordan was now giving the Iraqi opposition.

Divisions Among the Kurds

In other respects, the many factions of the Iraqi opposition were as ineffective as ever in hastening the end of the regime. Their main foothold within Iraq, in the Kurdish 'safe haven', was severely compromised by the continuing hostilities between the two major Iraqi Kurdish parties, the Kurdistan Democratic Party (KDP), led by Masoud Barzani, and the Patriotic Union of Kurdistan (PUK), led by Jalal Talabani. The peace accord, hammered out in Dublin in 1995 with US mediation, has still not been implemented. As a result, the parties maintained their respective positions in 1996, with the KDP occupying the strategic border crossing at Zakho and the PUK continuing to occupy Arbil, the administrative capital of the Kurdish region. This gave the KDP control of the very substantial dues levied on all through traffic to Iraq, while the PUK controlled nearly 70% of the region's population.

No effective government thus existed and the two parties administered their zones as they saw fit, seeking always to gain advantage over the other. The result was a series of armed clashes during the first half of 1996. Meanwhile, the Turkish government launched a number of military operations over the border against the forces of the Turkish Kurdistan Workers' Party (PKK) which had sought refuge in Iraqi Kurdistan. This led to a working relationship between the Turkish authorities and Barzani's KDP, which evidently believed that the continued link with Turkey was more important than any ties with the PKK. Equally, the Iranian government saw the 'safe haven' as providing a refuge for the Kurdish rebels of the Iranian KDP and stepped up its military operations against them. This culminated in a major cross-border raid by a mechanised brigade of the Iranian Revolutionary Guard in July, assisted by the PUK. The KDP, however, refused to cooperate and relations soured with Tehran. This shift undermined the position Iran had been seeking to occupy as an independent, powerful mediator in Kurdistan with good relations with both major parties. Instead, Iran seemed increasingly to ally itself with the PUK.

In August, as fighting between units of the KDP and the PUK escalated, this apparent regional alignment played its part. By the third week of August, the KDP was accusing the PUK of receiving direct Iranian military support for its attacks on KDP positions. As a result, Barzani warned that his own party would be obliged to ask the Iraqi government for military assistance. Barzani was clearly expecting some kind of US mediation to bring the fighting to an end, but when this was not forthcoming, he turned to Saddam Hussein who was only too willing to help. On 31 August, 30,000 Iraqi troops entered the Kurdish region and helped the KDP forces capture Arbil from the PUK. Inevitably, the Iraqi security services followed the troops and hunted down political opponents and army deserters, effectively breaking up the US-supported operations of the opposition Iraqi National Congress based in Arbil.

The sudden military operation launched by Baghdad took many by surprise, despite Barzani's warnings that he might be driven to seek Iraqi assistance. Concerned at the gains made by Baghdad and by the fact that this might herald an Iraqi reoccupation of the whole Kurdish region, the US launched a series of cruise-missile attacks on air-defence sites in southern Iraq on 3 and 4 September. At the same time, the US extended the southern no-fly zone northwards from the 32nd to the 33rd parallel. This now included a number of major Iraqi air bases and brought the zone close to the city of Baghdad itself. Iraq's response to these new restrictions was initially defiant as its remaining air-defence units tried to challenge US dominance of the skies. Given the scale of the US air superiority, this could only be a gesture of defiance. Even this retaliatory action was discontinued when the substantial reinforcement of the US Air Force in the region implied that the US was preparing for a major attack on Iraq (see map, p. 240).

By this stage, however, Saddam Hussein had achieved his purpose in Kurdistan. As he had predicted when fighting first erupted in 1994, the Kurds had quickly begged for Baghdad's assistance. The military operation to capture Arbil demonstrated to the world and to the Iraqi people that Saddam Hussein's regime was still a force to be reckoned with. He thereby warned Iran against meddling in Kurdish affairs and demonstrated to Turkey that it would be prudent to deal with him, as well as with the KDP.

The President's defiance of the US did him no harm domestically or regionally and it was clear that the US military response had not gained universal approval. On the contrary, only Kuwait and the UK seemed to offer the US their unreserved support. At the UN, a resolution condemning the Iraqi government was abandoned for lack of support at the Security Council. In assisting the KDP, Baghdad was not only responding to a clear request from a local party, but was also acting in an area still regarded as sovereign Iraqi territory, which caused many states to regard the US military response as inappropriate. France, for example, refused to cooperate in monitoring the newly extended no-fly zone, and later in the year it withdrew entirely from reconnaissance operations over northern Iraq.

In the event, the bulk of Iraqi forces were withdrawn within a week of capturing Arbil, leaving the KDP to pursue the demoralised PUK and to capture the latter's regional stronghold of Sulaimaniya on 9 September. This precipitated an exodus of thousands of refugees towards the Iranian border, but when it became clear that Iraqi government forces were not involved, many returned to their homes. Despite pleas from the PUK, and when it was clear that Baghdad had withdrawn, the US announced that it had no intention of intervening in what it regarded as a civil war between the Kurds. Instead, it looked for opportunities to bring the factions together. This was soon provided by Barzani, under pressure from Baghdad to sign an autonomy agreement with Saddam Hussein. To avoid so concrete a commitment, he turned instead to the US, pointing out that he had sought Baghdad's help out of desperation.

This gave the US the opportunity and incentive for more effective mediation when the next round of fighting between the KDP and the PUK broke out in October. Rallying his forces, Talabani launched a counter-attack on the overstretched KDP, recapturing Sulaimaniya and a number of other districts. Neither Iran nor Iraq appear to have been involved. Indeed, Talabani claimed he did not continue towards Arbil for fear of provoking another intervention by Baghdad. These reassurances led to intensive US mediation efforts which brought about a cease-fire by the end of the month.

During the following months a series of peace talks between the two parties was held in the Turkish capital, Ankara, under the auspices of the Turkish and US governments. Although the talks have not resolved the major issues, they have helped to monitor the cease-fire, reducing the likelihood that either party will feel compelled to call in aid from Iran or Iraq in the immediate future. Nevertheless, it was significant that in November and December the US airlifted some 5,000 workers and dependants from the Kurdish zone. This suggested that it had accepted the degree to which Iraqi security services had penetrated the zone, which made it hazardous for Kurdish workers associated with US projects to remain in the area. It was also taken by some as a weakening of the US commitment to continue *Operation Provide Comfort* which had hitherto effectively prevented the reassertion of Iraqi control over the whole of the north.

Iraq and the United Nations

Change was in any case in the offing since the Iraqi government and the United Nations finally reached agreement in May 1996 over the terms of UNSCR 986. This allows the Iraqi government to sell $2 billion-worth of oil every six months to pay for imports of foods, medicines and other humanitarian supplies. Following Saddam Hussein's decision early in 1996 that Iraq would be willing in principle to accept UNSCR 986, negotiations began in February to work out the exact terms of the 'oil-for-food' arrangements. These talks proceeded intermittently until agreement was reached on 20 May. During this period, the Iraqi government had been trying to evade the forms of supervision which would give the UN and its inspectorate detailed control of the allocation of funds earned from the licensed sale of Iraqi oil. In the event, Iraq was obliged to accept the UN Security Council's strict terms, minimising Baghdad's control of the funds and of the distribution of supplies.

Iraq's position was not much helped by its simultaneous embroilment in disputes with the UN Special Commission on Iraq (UNSCOM). Revelations in 1995 and other indicators had suggested that Iraq still possessed both biological weapons and a significant number of surface-to-surface missiles capable of carrying biological warheads. These fears seemed to be confirmed in March 1996 when the Iraqi authorities refused UNSCOM inspectors permission to enter a number of buildings suspected of housing details of Iraq's weapons programme. After a censure resolution from the

UN Security Council, the problem was solved, but predictably nothing was found. Iraq's obstruction had come a few days after the UN Security Council had voted to renew sanctions against Baghdad and may have been an act of defiance. The effect, however, was to delay talks on UNSCR 986.

In June, soon after agreement on the Resolution was reached, but when many details still needed to be worked out between the two sides before it could be implemented, Iraq tried a similar ploy. UN weapons inspectors were barred from entering three military sites, while Iraq levied accusations of espionage and threats to national security. This developed into the most serious stand-off for five years. The UN inspectors were finally obliged to leave the country, despite a unanimous UN Security Council condemnation of Iraq and the assertion that this was in violation of the 1991 Gulf War cease-fire agreement. The threat of military action behind this assertion led to a series of negotiations with the head of UNSCOM, Rolf Ekeus. As a result, Saddam Hussein agreed on 22 June to remove any future obstacles to inspection and to provide fuller documentation on Iraqi weapons programmes than had hitherto been the case. But this did not prevent the Iraqi authorities from blocking and delaying yet another team of UN inspectors in July, which brought Ekeus back to Baghdad in August to negotiate better access for his inspection teams.

By these actions the Iraqi government virtually guaranteed that the stringent UN sanctions regime would be renewed every two months. It also jeopardised the implementation of UNSCR 986, which Saddam Hussein believed presaged Iraq's full return to the world oil market. The costs he appeared willing to incur convinced the UN inspectorate of the scale of Iraq's concealed weapons capability and the importance of the role assigned to these weapons in long-term Iraqi strategy. Indeed, by the end of the year UNSCOM estimated that Iraq might still possess up to two dozen serviceable surface-to-surface missiles. This led to further disputes with Baghdad over its missile-destruction programme, resolved only when in February 1997 it was agreed that the parts of the destroyed missiles should be shipped to the US for detailed inspection to verify whether they were indeed key missile components.

The repeated confrontations between Iraq and UNSCOM meant that it was not until mid-August 1996 that the relevant UN committee formally accepted the guidelines for the 'oil-for-food' deal. However, all further progress was suspended for two months following the Iraqi government's intervention in Kurdistan. Apart from the United States' explicit hostility, Iraq's threat to Kuwait at the height of the crisis and the continuing unsettled condition of the Kurdish zone ensured that UNSCR 986 could not be implemented for some months. By November, however, talks had resumed. Agreement was at last reached at the end of the month on the oil-pricing formula and the way was finally open for Iraq to sell a limited amount of oil. On 10 December, Saddam Hussein switched on the pump to start the flow of oil through the pipeline to Turkey and, after some technical problems, Iraq once again began selling oil legally on the world market.

A Shaky Regime

It was important for Saddam Hussein's prestige to be seen re-opening Iraq's oil pipeline to the outside world, regardless of the fact that it was only possible because he had accepted a series of restrictive conditions imposed by the UN. He must continue to give the impression to his people that he is in control. This image suffered a blow when an attempt was made on 12 December on the life of his eldest son, Uday, who was shot and severely wounded in Baghdad. The would-be assassins escaped, but the government blamed the underground Shi'i opposition group Al-Da'wa which also claimed responsibility for the attempt. Nevertheless, there are strong suspicions that those responsible may in fact have had more personal grievances against Uday. His brutality and corruption are notorious in Iraq and there are many, including some of his own kinsmen, who have reason to wish him dead.

Regardless of the identity of the attackers, it was significant enough that they obtained sensitive information on Uday's whereabouts and movements. In response, Saddam Hussein ordered purges not only of the forces responsible for protecting the President's family, but also more generally of army officers and officials. These extended well into January 1997 and touched members of clans hitherto close to Saddam Hussein. Determined to pre-empt any moves by others to exploit differences within the family and the clans that surround him, he placed his second son, Qusay, in charge of the interrogations. However, in late February, there were unconfirmed reports of a failed assassination attempt against Qusay in Baghdad, which may have been part of the same pattern of enmity and revenge.

The politics of the clans at the heart of the Iraqi regime are more obscure than the activities of the many parties that constitute Iraq's declared opposition. Nevertheless, it may be precisely such intimate tensions and rivalries that will eventually bring about Saddam Hussein's downfall. In another configuration, they may also form the basis of his successor's power. This could be encouraging for many in the region who wish to see Saddam Hussein overthrown, but who fear some of the imagined consequences. Yet the experiences of the past few years would suggest that the very intimacy of the relationships at the heart of power in Iraq can lend them a resilience and an imperviousness to outside manipulation which make the outcome of the tensions that do arise difficult to predict.

These often obscure manoeuvres demand constant vigilance, but Saddam Hussein has shown himself to be a master of the game. It is against this background that Iraq calculates its moves in international politics. As far as UNSCR 986 is concerned, Saddam Hussein probably believes that Iraq's modest return to the world oil market signifies the beginning of the end of the international sanctions imposed against it since 1990. This may well be why he finally accepted a deal which he had hitherto characterised as a gross infringement of Iraqi sovereignty.

He may be right in his estimation. There have been signs that many states are willing to contemplate Iraq's rehabilitation, even under the present regime. Nevertheless, as the actions of his government during the past year demonstrate, this is not Saddam Hussein's only concern. Iraq's weapons programme clearly continues to occupy a position of such importance in his thinking that he is willing to run the risk of the UN Security Council unanimously renewing sanctions for the foreseeable future. Equally important, as his actions in Kurdistan have shown, he is willing to place the economic rehabilitation of Iraq in jeopardy in order to maintain the credibility, deterrent power and standing of his regime within Iraq. It is, after all, on this that his survival largely depends.

Asia

The politician's adage that 'all politics are local' was truer than ever for Asia in 1996. The twist was that local politics had major implications for regional and wider international relations, and as a result a new balance of power in Asia began to emerge, albeit slowly.

The most important domestic politics was that of China, and it was China's actions that provoked the new balance of power. The much-anticipated death of Deng Xiaoping in February 1997 had little immediate impact on the country's domestic or external relations. But the fact that the struggle for the fate of China's reforms could now be openly fought will soon lead to a clearer sense of whether the country really is re-emerging as Asia's dominant power. China's domestic uncertainty and unsteady rise is a key factor driving Japan to reform itself and adjust its own foreign policy. Closer US–Japan and Russo-Japanese relations in 1996 suggested that Japan feared it might become vulnerable if China became more assertive.

Korean affairs remained fixated on peninsular issues. Domestic scandals and unrest in South Korea paled beside the North's dire economic straits and lethal factional politics. Elsewhere in East Asia many countries were concerned about the sustainability of their economic growth rates. In South Asia, elections in Pakistan and India were self-absorbing, but remarkably India managed to sustain its commitment to economic reform and greater interdependence with the outside world. India's determinedly open outlook, coupled with rising concern in East Asia about sustainable growth and a rising China, meant that 1997 began with the stirrings of what might become a regional Concert of Powers.

A More Assertive China

In the early 1990s, it became clear that China was a rising power. By the mid-1990s, there was considerable evidence that China planned to use its new-found influence in an assertive way. The second half of the decade will demonstrate whether the outside world is prepared to restrain as well as engage an assertive China. The signs are mixed. In 1996, when confronted by a firm US response to its policy on Taiwan, by a phalanx of international pressure to come to terms on the Comprehensive Test Ban Treaty (CTBT), and even by a more united Association of South East Asian Nations (ASEAN) concerned about Chinese actions in the South China Sea, Beijing carefully backed down. In 1997, the key gauge of China's assertiveness and of the external constraints on its actions will be the retrocession of Hong Kong to China on 1 July.

China's assertiveness, and the ambiguity of other states' responses to its power, stem from the uncertainties surrounding China's long-term economic and political prospects. With the volatile mixture of a strong economy and a struggling political system, the country veers from excessive confidence to paranoia about its ability to prosper free of foreign constraint. Similarly, external powers are anxious to benefit from the growing economy, but also hope in the short term to restrain China and make it a more amenable international partner.

Reactions to the death of China's paramount leader, Deng Xiaoping, on 19 February 1997 provided graphic evidence of the uncertainties in both China's political system and the outside world's attitudes towards it. The implications of these uncertainties, however, need to be seen in a longer-term perspective. Deng's death may well have a significant effect on China's economic and political prospects, and on the outside world's reactions to China, but these issues were at stake before Deng died and will only be resolved well after his ashes have been scattered.

Keeping the Economic Engine Running

Deng Xiaoping's greatest legacy to the Chinese people – one-fifth of the total world population – was economic reform. China's spectacular economic success remains the key to its power. As long as economic growth can be sustained, there will probably be sufficient wealth to buy off domestic protest and to keep foreigners interested in maintaining good relations with Beijing. In recent years, the Chinese economy has more often run the risk of overheating than of the engine grinding to a halt.

China has averaged nearly 10% growth since economic reforms began in 1978. The key to this success has been the decentralisation of the economy. But paradoxically, the very success of decentralisation makes it difficult for central government to manage the economy. In 1996, the leadership in Beijing seemed to have abandoned many aspects of its struggle with provincial and local government and opted for a more explicitly coordinating role. For much of the previous 7–8 years, central planners had struggled to regain control of the economy after allowing entrepreneurs, provincial leaders, township and village enterprises, and even the children of senior leaders, to help China grow by pursuing diverse schemes for economic development. In the early 1990s, when Beijing struggled to grasp the levers of economic power, it discovered that decentralisation had gone too far and that it could no longer achieve cooperation through coercion.

Thus in 1996 Beijing, perhaps only temporarily, offered some concessions in return for cooperation. In Communist Party-speak, this became known as 'bartering tolerance for obedience'. The challenge now facing the various parts of the decentralised economic system is how to institutionalise these relations to create more orderly and effective government. This 'repositioning' of central government may very well make China's economic policy more effective.

After failing for nearly four years to decelerate, China's economic growth rate did stabilise at about 10% (10.2% in 1995 and 9.8% in 1996). Even more spectacular was the cut in the consumer price index from 17% in 1995 to 8.3% in 1996. Foreign-currency reserves in January 1997 were the world's second largest at $105 billion. China's foreign trade surplus for 1996 was a healthy $12.4bn, and total foreign investment in 1996 reached a record high of $42bn, although contracted foreign investment was down by a third to $70bn from its peak of $104bn in 1993. However, China felt confident enough to succumb to long-standing pressure from the World Bank and the International Monetary Fund (IMF) and agreed to convert the *renminbi* on the current account on 1 December 1996.

Of course, even Chinese officials take Chinese statistics with a pinch of salt, but there was a clear downward trend in the economy – what officials called a 'soft landing'. Its extent was, however, a matter of dispute, and by the last quarter of 1996 there were signs that the landing was much harder than had first been thought. Precisely because Chinese officials feared the consequences of slow economic growth more than hyper-growth, rapid efforts were made to reflate the economy in the first quarter of 1997, even at the risk of re-stoking inflation. Given the audacity and scale of China's economic and social reforms, this was probably the only sensible tactic, but it could hardly be called 'an economic strategy'. The economy remained far too chaotic and decentralised for there to be any coherent strategy and Beijing's grudging acceptance of the realities of a decentralised economy has left it too divided to formulate any serious long-term policy.

Leadership, 'Followership' and Indecision

Chinese officials habitually reassure the outside world that their leadership is stable and well entrenched until the moment it changes. Then a new 'long-term' status quo is declared. Despite their regular exhortations to trust them, Chinese officials are becoming used to not being believed abroad. They are certainly not believed at home. When Deng Xiaoping died on 19 February 1997 aged ninety-two, nothing much changed.

Since Deng's efforts in 1992 while touring the south of the country to revitalise his faltering reform programme, he had ceased to be an active player in Chinese politics. Although the collective leadership saw the virtue of containing their disputes and avoiding difficult decisions, there has been little doubt in the five years since 1992 that China's leaders are divided on important issues of economic and social policy. Now that Deng has died, these debates are likely to become more fierce. In the past, the Chinese leadership has tended to change as a result of such debates, and

Jiang Zemin

Li Peng

Qiao Shi

China's Major Leadership Contenders

Zhu Rongji

Li Ruihuan

there are good reasons to believe that China's current Party leader and President, Jiang Zemin, will have to fight hard to retain his position. He will have to prove that unlike Mao Zedong's designated successor, Hua Guofeng, he can grow beyond being a compromise candidate in a collective leadership and become a genuine leader.

As with previous leadership transitions in China, the hidden reality has to be teased out of esoteric leadership debates, but the current line-up does not simply disagree over faster or slower reform. China has always been, and is now more than ever, a highly complex society and polity. Particularly in the uncertain period before Deng's death, it was often safer for leaders to pretend to be followers and not expose a vulnerable neck to the chop of factional politics.

In the year before Deng's death it was clear that some in the leadership recognised that China's economic reforms could not just be sustained, they needed to be driven forward. These forces were said to include Vice-Premier Zhu Rongji – described as 'China's Gorbachev' before the analogy became more of a liability than an accolade – and the somewhat younger former Mayor of Tianjin, Li Ruihuan. Li Ruihuan is touted as a possible successor to Premier Li Peng, who must retire in March 1998. Qiao Shi, Chairman of the Standing Committee of the National People's Congress (the Chinese parliament), has been seen as a possible more liberal successor to Jiang Zemin in a post-Deng shake-out. In anticipation of the 15th Party Congress due in October 1997, factional struggles in the leadership have clearly begun. Initially, there were signs that Jiang Zemin was drifting towards the liberal wing of the Party in order to build a broader coalition against conservative forces, but his credentials as a committed reformer are open to doubt. His recent conversion to reform could be yet another policy change by the politician known as the 'weathervane'. The complex leadership struggle is likely to go through many such twists and uncertainties before the future of Chinese reform becomes clear.

Longer-term confidence about the sustainability and expansion of the reform programme comes from the fact that advocates of faster reform are said to include many younger, middle-ranking officials. Reformers gain much support from provincial leaders who have benefited from the decentralised economy. They are also supported by large parts of the collective (but not state-owned) sectors of the economy where most participants want even more freedom. They are even said to have the support of some state-owned industries which think they can prosper through privatisation. Those parts of the armed forces that are busy making money or wish for a more capable economy to support the high technology necessary for a revolution in military affairs are also said to be sympathetic to such ideas. Clearly this is a complex coalition of interests.

The advocates of faster reform maintained a low profile for much of the years before Deng's death because of the problems caused by the currently restrained reforms. The state-owned sectors of the economy are haemorrhaging money. With no social security system, reform of state-owned industry has stalled and social instability is increasing. Corruption is rampant – so much so that China is now ranked among the top three most corrupt nations. The municipal leadership under Chen Xitong in Beijing was so riddled with corruption that heads, including his, have had to roll. Zhu Rongji's prestige also suffered when he was unable to control the overheated economy in 1994–96 and central government essentially had to give way to provincial and local power.

The epitome of the conservative critique is the elderly Deng Liqun, but this view is also manifest in the front-line leadership of others such as Premier Li Peng. Deng Liqun (dubbed 'the lesser Deng') supervised a 10,000-character critique of Deng Xiaoping's reforms in September 1995, and a pseudonymous 'Fan Liqun' responded with an unpublished 80,000-character defence of the reformist cause. At the annual retreat of senior leaders at Beidaihe in August 1996, both arguments were apparently presented, and Jiang chose a strategy of 'peaceful coexistence' in an attempt to dilute the dispute.

The conservative forces in the leadership are best described as cautious reformers, for many are simply worried about uncontrolled reform and are not necessarily in favour of a return to old communist values. Some, like Deng Liqun are, in Communist Party jargon, said to favour a return to 'taking class struggle as the key link' rather than 'taking central construction as the main task'. But the younger, more nationalist conservatives want more authoritarian, but not necessarily more socialist, rule. Thus it can be argued that the core reforms are irreversible. Current experiments with very limited forms of democracy at village level, as well as widespread decentralisation, may be undone by a conservative resurgence, but the greater role for market forces seems likely to endure. Nevertheless, it is misleading to talk of irreversibility when the real challenge is how to proceed, not whether current reforms will unravel. Precisely because the leadership agrees that China's reforms cannot simply be more of the same, the more liberal forces fear that the conservative critique will lead to an effective reversal of key reforms.

Lesser Deng's 1995 manifesto was the focus of much heated debate in 1996. Jiang Zemin thought it prudent at the 6th plenum of the 14th Party Congress in October 1996 to take the credit for a stiffer campaign against corruption and a new campaign to 'build spiritual civilisation' and promote 'ideological, moral and cultural progress'. The campaigns led to a crack-down on more liberal social forces and (mostly fruitless) attempts to reassert greater control over the media, especially in coastal China. Although the conservatives were not actually winning the debate, no one was prepared to fight them to the end before Deng's death. Now that Deng has died, these debates can be translated into real struggle and changes in policy. Key signposts in this process will be the changes in senior leaders' positions on the speed and direction of reform in the run-up to the annual summer leadership retreat at Beidaihe. The most important struggle will centre on the personnel and policies that will be presented at the 15th Party Congress in October. The fact that Jiang Zemin may survive as Party leader only by becoming more committed to reform suggests just how far observers should focus on changes of policy rather than changes in leadership.

In this atmosphere of complex contending forces, it was safest and easiest when Deng was alive to be a leader skilled in 'followership' and indecision. Hence the prominence of President Jiang Zemin. Various leadership changes in 1996 were accepted as signs of Jiang's growing hold on power and the Party's nomenclature. Now that Deng has died, Jiang may find that followership is a liability and only leadership will secure his future. Jiang can do little to change his grey character, but because, unlike Hua Guofeng, he has no Deng Xiaoping at his heels, he may survive for some time. Yet if China really wants to commit itself to rapid reform, it is more likely to choose a leader like Qiao Shi, or more daringly and perhaps in the longer run, Li Ruihuan, to drive the reforms forward. Jiang's long-term survival may signal continuing muddle and compromise at the top which may put the reforms at risk. The rise of Li Peng would signal serious trouble for the reform programme. In short, Deng's death means that China can now seriously confront the tough choices about reforms that it has been avoiding for many years. Anything short of renewed and sustained reform will put Deng's legacy at risk.

China Pushes ... But Only Some Push Back

The domestic choices China makes have a major impact on its foreign policy. China's first foreign-policy act in 1996 was a major assertion of its right to order the lives of the Taiwanese people. China challenged Taiwan, and indeed the wider world, by attempting to influence the outcome of the March 1996 Taiwanese election through military threats. When China closed air and sea lanes around Taiwan, the United States responded with a brusque and efficient demonstration of military deterrence by deploying two aircraft carrier battle-groups.

It is doubtful that China ever planned to take Taiwanese territory, but the intention to intimidate was real and required countervailing deterrence. China learned several lessons from this episode. Not only did its armed forces not perform particularly well in their military exercises (they had trouble operating in bad weather), but it also became obvious just how far China still is from being able to cope with superior US military might. US forces could operate at a considerably longer distance and yet have more real-time data and effective military power than China. China's armed forces were taught the same lesson they had learned from the 1991 Gulf War – they are a generation behind the revolution in military affairs. Because the crisis was so close to home and several years after the Gulf War, it was all the more shocking that China remained an inferior power.

As Chinese security planners went back to the drawing board for a longer-term and more sustained effort to compete militarily, Beijing adopted a more flexible political strategy. It sought and eventually received the promise of a presidential summit with the United States in 1997. US efforts to engage China in cooperative relations – for example, a pledge not to target each other with nuclear weapons – was rejected during 1996 by a still-smarting and innovation-challenged Chinese leadership. Given the debates in Beijing and the failures of its Taiwan policy, such sullenness was not surprising, even if it was still poor policy. The new US foreign-policy team settling in to office in early 1997 seemed committed to dialogue with China, but there were signs that they had learned the need for firmness in dealing with Beijing that had so obviously been lacking in the first part of President Bill Clinton's first-term China policy. Even during new Secretary of State Madeleine Albright's visit to Beijing just after Deng's death, there was no shying away from difficult issues such as trade disputes and human rights. In March 1997, as reports appeared about Chinese funding for American politicians, it was clear that 1997 might be just as difficult a year as 1996 was for Sino-US relations.

China's foreign-policy strategy was more innovative the further it moved from what it had clearly identified as the major long-term challenge – the United States. In classic 'united-front' tactics from the revolutionary era, China sought greater unity with forces that would help it contain the power of the 'primary contradiction'. The 'united front' strategy bears many similarities to what China decried as the US attempt to build a containment coalition against it; in truth, both powers were manoeuvring in similar ways in the fluid regional balance of power.

China's strategy was successful in that in December 1996 it finally persuaded South Africa to sever diplomatic ties with Taiwan. South Africa was the last large country to try to retain formal relations with both Taiwan and China. China also began to use bully tactics at the United Nations Security Council – blocking resolutions and action on Haiti and Guatemala because both countries had good relations with Taiwan.

This exercise of crude power was in contrast to, but just as successful as, its more subtle change of policy towards South-east Asian states. China clearly understood that if it was to find allies and undermine the US strategy of being the regional balance, its best chance lay in persuading ASEAN countries that the United States was unreliable and that their future lay in hitching their fortunes to China's future. Hence there was a clear change in Chinese policy towards the ASEAN Regional Forum (ARF) in 1996. Beijing agreed to become an active participant in ARF and Council for Security and Cooperation in the Asia Pacific (CSCAP) activities in order to make the case that China, and not the United States, was the country most genuinely committed to regional multilateralism. A similar calculation lay behind China's warm endorsement of the first Asia–Europe Meeting (ASEM) in 1996 – an event Beijing designated one of the top ten 'world news event in 1996'. By welcoming the fox into the chicken coop, however, ASEAN may find that the ARF will be unable to deal with any difficult security issue in which China was a primary protagonist. Similarly, China's membership of ASEM may prevent its summit from tackling tough economic, political and security issues.

Although the policy towards both ASEAN and ASEM meant that in 1996 China moved closer to ASEAN and European Union countries, relations with Japan took a turn for the worse. Japan's closer security links with the United States emerged as a major bone of contention between China and Japan. When Prime Minister Ryutaro Hashimoto spoke in Singapore in January 1997 in coded terms of regional concerns about China, Beijing saw this as yet another attempt by Japan, the United States

Table 1 *China's Recent Acquisitions of Foreign Weapons and Technology*

Classification	System	Supplier	Order date	Quantity	Delivery date	Deliveries to date	Comment and technology-transfer details
satellite	See comment	Various	1996	9	2000	4	Access to remote-sensing satellites
submarine	*Kilo*-class	Russia	1992	4	1994	2	2 *Kilo*-type 877 delivered in 1994 and 1995 (in UN Register of Conventional Arms). Order for two type 636 for delivery in 1997
destroyer	*Sovremenny*-class	Russia	1996	2	1999	0	Equipped with SS-N-22 *Sunburn* anti-ship cruise missiles
fighter/bomber	SU-27	Russia	1990	50	1992	50	Two batches of 26 (in 1992 UN Register of Conventional Arms) and 24 delivered in 1996, equipped with Russian AA-10 air-to-air missile
fighter/bomber	SU-27	Russia	1996	150			Agreement to produce under licence up to 150 modernised SU-27 models
transport aircraft	Il-76	Russia	1992	15	1994	15	Assembled in Uzbekistan. Deliveries into 1995
airborne early warning modification	Il-76	Israel	1996	1			Modification with Israeli cooperation using *Phalcon* radar. Russia delaying delivery of Il-76
maritime patrol aircraft modification	Y-8	UK	1996	8			Modification with UK cooperation to install *Searchwater* radar
attack helicopter	HJ-8/Z-9	France	1980	50	1988	50	Army and Navy attack variants exist. Part of initial batch of 50 ended in 1992. Production continues
transport helicopter	Mi-17	Russia	1992	28	1994	28	
light utility helicopter	EC-120	France	1990		1998		Civil helicopter with military potential, developed in collaboration with Franco-German Eurocopter and Singapore Aerospace
surface-to-air missile	SA-10/S-300	Russia	1992	500–900	1993	144+	China reportedly building under licence

Note: The contract values of the listed transfers are estimated to be around $5bn
Source: IISS, London

Table 2 *Major Chinese Weapons Programmes Incorporating Foreign Technology in 1997*

Classification	System	Order date	Quantity	Delivery date	Deliveries to date	Foreign-technology content
inter-continental ballistic missile	DF-31/41	1997		1998		Russian/Ukrainian SS-18/19 ICBM technology
submarine-launched ballistic missile	JL-2	1985		2003		Similar to DF-31 series
intermediate-range ballistic missile	DF-21	1983	50	1987	50	
short-range ballistic missile	DF-11	1984	200+	1992	200+	Export variant known as M-11
short-range ballistic missile	DF-15	1984	400+	1991	400+	Export variant known as M-9
nuclear ballistic missile submarine	Type 094	1985		2003		Entry into service after 2000
nuclear submarine	Type 093	1985		2001		With Russian design. Entry into service after 2000
satellite	DFH-3	1988	2	1994	2	Dual-use communications satellite.
satellite	remote sensing	1988	2	1996	1	Development programme with Brazil
satellite	FSW-2	1975	3	1992	3	Designed for civil and military applications.
main battle tank	Type 90	1990		1997		Developed from Type 85IIM. Trials in 1996.
main battle tank	Type 85 IIM	1995	1,200	1989	1,200	
armoured personnel carrier	Type 90	1990	2,000+	1995	400	Family of 12 AFVs. Production started in 1995.
air-capable ship		1996	2	2000		Accelerated programme reported late 1996.
submarine	*Ming*-class	1992	1	1996	1	Equipped with US satellite navigation
submarine	*Song*-class	1985	3	1995	1	German engine. Testing YJ-82 anti-ship cruise missile.
destroyer	*Luda* III-class	1988	2	1993	1	With French and Italian combat systems.
destroyer	*Luhu*-class	1985	3	1994	1	US engines, French and Italian helicopters, electronics and weapons
frigate	*Jiangwei*-class	1988	6	1991	4	Four built to date. Two ordered, apparently delayed
fighter/bomber	F-7	1961	2,000+	1961	2,000+	Licence-produced Russian MiG-21.
fighter/bomber	F-10	1989	300	2003		Russian engines and Israeli design and electronics
fighter/bomber	FC-1	1990		2000		Former Russian MiG-29 with European and Israeli avionics.
fighter/bomber	F-8II(M)	1994	200	1997		Former Russian MiG-23. 1996 upgrade with Russia
trainer	K-8	1987	300	1994	34	With Pakistan, US engines (subject to US embargo)
tanker	Y-8	1992	8	1994	8	Based on Russian AN-12

Source: IISS, London

and Australia to pressure potential allies in ASEAN (perhaps Indonesia and the Philippines) to take a tougher line against China. In late 1996, the Sino-Philippines dispute over the South China Sea revived when Beijing protested at the extension of a landing strip on a Philippine-held island.

Japan's growing concerns about Chinese intentions were highlighted when the dispute over the Senkaku/Diaoyu islands flared up again in September 1996. These Japanese-held islands in the Ryukyu chain were the target of ethnic Chinese nationalists in Taiwan and Hong Kong who protested against the establishment of a lighthouse (described by some in the region as a 'lamp post') on the islands by Japanese nationalists. What was curious, and in great-power terms encouraging about the dispute, was the way in which Japanese and Chinese officials tried to play it down. Both feared the risks of uncontrolled nationalism, and Beijing in particular suspected that once protesters took to China's streets, their grievances could turn more inward looking.

Although the tension soon subsided, the residual fear of the increased power of nationalism remained in East Asia. This concern was yet another part of the regional states' subtly coded exploratory strategic efforts to find a new way to cope with rising and falling powers in their region. Inevitably, Russia was drawn in to the process. Japan and Russia began a series of military-to-military consultations, despite their continuing failure to agree on disputed territorial issues. At the same time, Russia and China further developed their so-called 'strategic partnership'. Although Russia did transfer significant weapons systems to China in 1996 (see Tables 1 and 2), the relationship was far less strategic and certainly far less of a partnership than many Chinese liked to pretend. Russia clearly felt there was money to be made from arms sales and that in the short term it was worth showing the West that Russia had friends in other places. Moscow was not naive about China's long-term power and intentions, but in the short term improved Sino-Russian relations were seen as a way for Russia to return to the mainstream of East Asian international relations. For Beijing, exaggerating the extent of the strategic partnership was clearly helpful in counter-containing the United States.

In 1996 there were two high-level Sino-Russian meetings. On 26 April in Shanghai, the Presidents of China, Russia, Kazakstan, Kyrgyzstan and Tajikistan signed an agreement on military confidence-building measures in the border region. This far-reaching accord – the most comprehensive arms-control agreement in East Asia since 1945 – restricted deployments and exercises in the border region and, uniquely for Asians who claim they are culturally indisposed to reach such legal accords, the pact was lengthy and legalistic. The second high-level meeting between Prime Minister Li Peng and ailing Russian President Boris Yeltsin in Moscow on 27 December produced no major agreement, but talk of an important 'strategic partnership' was given great prominence (see maps, pp. 250–51).

The United States' cautious reaction to Sino-Russian relations reflected Washington's overall indecision about how to define the 'Chinese challenge'. No longer was China merely an emotional issue related to human rights – although that aspect still remained. Now China was overtaking Japan as the country with the largest trade surplus with the United States and the country that necessitated the most difficult trade negotiations. The attitude China had adopted before agreeing in summer 1996 to join a CTBT, as well as its insistence on its right to proliferate weapons and nuclear materials to rogue states in the Middle East and to an unstable Pakistan, created a major problem for proponents of non-proliferation. China even tried to prevent the Disney Corporation from distributing a film about Tibet, and increased its efforts to intimidate the US press corps in Beijing.

As time went on, China seemed poised to present a challenge difficult to surmount. But as the CTBT, trade issues and even the Disney case made plain, it was possible to force China to make concessions and constrain its policies. Such concessions and constraints were not achieved by accepting China's definition of what 'engagement' should be, but rather through a firm determination to pursue individual interests and force China to compromise. As the CTBT issue showed very clearly, China was prepared to defend its own interests, but would agree to concessions when it became the only country to delay an accord. And as the various trade disputes with China showed, Beijing needed access to US (and other Western) markets so badly that it could not allow a trade war to break out. China was being forced to accept the constraints of interdependence.

The Fate of Hong Kong

The extent of China's assertiveness and the nature of other countries' responses will be tested in 1997 in a unique laboratory. The transfer of Hong Kong to China is an unusual episode in many ways – not least because it is a major international event whose date has been known ever since the Sino-British Joint Declaration in 1984. As a result, more than three thousand foreign journalists are expected to congregate in Hong Kong to watch the transfer of sovereignty. But the transfer is more important than diplomatic theatre; it is the last significant chapter in the history of European colonialism. But this is de-colonisation as never seen before. Rather than transferring sovereignty to the people, sovereignty is being transferred, against the wishes of a majority of the people, to the rulers of a poorer and more authoritarian country.

Never has a colony been as rich at the point of transfer, but the change of rulers can hardly be represented as adhering to the principles of self-determination. The UK and the people of Hong Kong had no choice in the matter; China controls the water, food and power supply to Hong Kong and therefore can impose its will. Given such realities, all parties are making the best of what has to be. The Hong Kong people are well on their way to establishing ever-closer economic ties with China, self-censoring their views and their media, and keeping their fingers crossed that China will exercise a light touch. If everyone remains calm, the optimists argue, China may recognise that it is best to maintain as much of the island's political and economic system as possible. In the long run, as China grows richer and freer, it may grow ever more tolerant of Hong Kong.

China feeds such optimism through its notion that Hong Kong can be part of one country, but retain its own separate political and economic system. Deng Xiaoping's slogan 'one country, two systems' is clever, but so vague as to be vacuous. It all depends on who will decide what features of Hong Kong's system can be retained. China formally pledges only to take control of foreign and security policy. Curiously for a place built on an export economy, foreign policy is supposed to exclude all matters of international trade and finance. Not so curiously for a Communist Party-run state, China claims that foreign and security policy include anything to do with sovereignty. Because national security requires internal public security, that too is a matter for Beijing. Apparently China will even revise Hong Kong's history textbooks. The result is much uncertainty about how China will behave after the handover. The fact that Deng died months before Hong Kong reverts to Chinese rule offers some hope that the key element of uncertainty – where will China go after Deng – may be reduced.

As 1 July 1997 approaches, the focus has been on smoothing out the transition process. The Legislative Council elected in April 1995 under political reforms established by Governor Chris Patten will be replaced by a provisional legislature chosen by a committee of Hong Kong people selected by Beijing. In November 1996, this committee also picked Tung Chee Hwa as the chief executive to replace Governor Patten. A prominent Hong Kong tycoon, Tung's company was bailed out by a Chinese-run enterprise and by March 1997 he had not criticised a single Chinese policy. Thus, when China announced in January 1997 that it would reverse various civil liberties introduced in Hong Kong in recent years, Tung publicly offered China his clear support, support he gave even more fulsomely in private. According to opinion polls, his vision of a Singapore-style Hong Kong was supported by a mere 17% of its people.

As China and its 'united front' supporters in Hong Kong clawed back recent reforms, a number of Western governments (the UK, Canada and Australia) protested and US President Bill Clinton called Chinese behaviour a 'litmus test' of its broader preparedness to accept the constraints of an open and interdependent world. China rebuked the outside world for protesting at its handling of Hong Kong, claiming that this was an internal matter of Chinese sovereignty.

These exchanges demonstrated that Hong Kong's system would be closer to that of China's than suggested by the 'one country, two systems' slogan. The spats fuelled concerns that China would impose its own system's lack of respect for the rule of justice and would tolerate mainland-levels of corruption. A self-censored Hong Kong press would hold no one to account. But Beijing and the people of the island calculated that no one was prepared to constrain Chinese behaviour.

Hong Kong had been granted a special status in hundreds of international accords as part of the outside world's support for the notion that it would have a genuinely separate system from the mainland. Should any of those benefits be withdrawn, the first people to be hurt would be its inhabitants. It is unlikely that any Chinese action in 1997 will provoke the United States and other Western countries to impose penalties on Beijing.

Nevertheless, the importance of the transition in Hong Kong is not just about the fate of six million of the richest people in Asia. It is also about what the process reveals about what China will be like when it grows strong. At several points in 1996, China's assertiveness was met by a firm response from the outside world. Although China was warned that a similar response was likely if it did not allow Hong Kong to retain its own distinctive system free of Beijing's control, the evidence so far is that China is prepared to be particularly assertive over Hong Kong, and the outside world feels it has few cards to play against such assertiveness. In the short term, China could no doubt succeed in doing as it pleases regarding Hong Kong. But in the longer term, China is planting and feeding seeds of worry about the implications of a rising China. Just as the massacre in Tiananmen Square in June 1989 had a negative impact on China's reputation well beyond the intrinsic horror of the event, so a heavy-handed Hong Kong policy will do long-term damage to China.

Japan: Preparing for Change

It was not quite business as usual in Japan in 1996. On the surface, as some of the older members of the ruling Liberal Democratic Party (LDP) appeared to believe, it looked like a return to the 'good old days'. The LDP's restoration to the premiership as part of a coalition government with the Social Democratic Party of Japan (SDPJ) and the *Sakigake* (Harbinger) party in January 1996 led almost inexorably to a sole government role after elections to the Lower House in October. For their part, the other coalition parties did little more than display a penchant for self-destruction.

But, as Prime Minister Ryutaro Hashimoto himself appreciated, looks could be deceptive. For one thing, the economy was still not performing well: a brief resurgence in early 1996 soon petered out, leaving business leaders and the public calling for fundamental change rather than short-lived cash infusions to regain vitality and competitiveness. At the same time, one of the pillars of Japan's post-war political and economic system –

the bureaucracy – came under increasing criticism for incompetence and dishonesty from politicians as well as from the people.

In foreign policy too, although the US–Japan alliance was dealt a blow by the uproar on Okinawa over the rape of a young girl by three US soldiers on 4 September 1995, Hashimoto and US President Bill Clinton were not only able to reaffirm the 1960 Japan–US Security Treaty in April 1996, but also to strengthen it by enlarging the areas of cooperation. Japan's relations with its Asian neighbours suffered a series of problems, primarily over territorial issues, and the seizure of hostages in late December 1996 at a Christmas party at the Japanese embassy in Lima, Peru, left the LDP government feeling helpless.

Thus, although Hashimoto entered 1997 as, arguably, the strongest LDP leader since Yasuhiro Nakasone in the mid-1980s, he had little reason to feel complacent. He has indicated that he is serious about introducing significant political and economic change in Japan. Yet he faces serious domestic and foreign-policy challenges that will require decisive action of a kind that has not for many years been the hallmark of the party he leads. Unless Hashimoto can translate his ideals into effective political leadership in 1997, the much-talked-about political and economic changes that could help solve some of Japan's underlying problems will remain only talk.

More Musical Chairs

In the first half of 1996, the LDP and its coalition partners, the SDPJ and *Sakigake*, encountered two major problems which provoked domestic controversy: the need to bail out the bankrupt *jusen* (housing-loan corporations); and renewing the leases for US military bases in Okinawa.

At the end of January, the government put forward a plan to liquidate the debt-ridden *jusen*, but only by presenting the tax-payers with a bill expected eventually to reach 1 trillion yen (US$8.8bn). Although the opposition parties tried to capitalise on public dissatisfaction, and instigated a series of stormy debates in the Diet, they were unable to stop the bill's passage through the Lower House in June. The Okinawa issue was more protracted, but the Clinton administration's willingness in April, however grudging at first, to consider reducing the US troop presence there helped to defuse some of the tension.

By the summer, therefore, Hashimoto began to consider calling an election. His coalition partners, which both feared for their survival, initially stalled, but finally gave way. The Lower House election on 20 October 1996 was the first since the LDP lost power in 1993 and also the first under a new electoral system intended to correct the past corruption scandals and 'money politics' abuses. The old multi-member constituencies were abolished and replaced by 300 seats directly elected in single-member constituencies and 200 seats allocated by proportional representation in 11 large regional blocs.

Japan's largest opposition party at the time of the election was the New Frontier Party (NFP or *Shinshinto*), headed by Ichiro Ozawa, like Hashimoto a younger-generation leader. Even within his own party, however, Ozawa's abrasive style had its critics, most notably former Prime Ministers such as Tsutomu Hata and Morihiro Hosokawa. A more serious problem for the NFP, however, was the sudden emergence, barely a month before the election, of the new Democratic Party (DP), jointly headed by two ex-*Sakigake* 'reformists', Naoto Kan and Yukio Hatoyama. This new party drew on disaffected politicians from almost all the old parties.

The Japanese public, which had been looking for something new and vibrant, was alienated by the choices on offer, not least because the political platforms were almost uniformly bland. Administrative reform and deregulation headed most parties' lists of policy priorities, but the proposals for carrying out the policies were vague at best. As a result, the voter turnout, at just under 60%, was the lowest ever recorded for a Lower House election in the post-war period.

Although the election result fell short of the LDP's ideal, an absolute majority, the Liberal Democrats derived some satisfaction from it. By improving its standing from 206 to 239 seats out of 500 it reinforced its position as the largest party in Japan. The NFP, by contrast, was disappointed to gain only 156 seats. Although the voter-mobilisation network of the former *Komeito* party – the political wing of the Soka Gakai Buddhist lay organisation and a constituent part of the NFP – was effective, Ozawa's reputation as a political fixer seems to have worked against his party. In addition, the NFP's glib promise of tax cuts, without a clear explanation of how these would be financed, did not convince voters. The Democratic Party gained 52 seats, a good result for a new party which had had little time to prepare for the election, but still short of its leaders' expectations. The insufficient clarity with which they had presented their policies to the voters had clearly mitigated their appeal.

While the opposition parties had not fared as well as they had hoped in the election, they hardly suffered compared to the LDP's former coalition parties. The SDPJ, which had tried to reinvent itself by reinstalling Takako Doi – an 'old-school' socialist – as leader, ended up with a mere 15 seats, half its previous total. For its part, *Sakigake* all but disappeared, reduced to a miserable two representatives. The only other party to gain ground in the election was the maverick Japan Communist Party, which at least had the advantage of clearly articulated policies. The voters were evidently weary both of the cycle of party splintering and amalgamation, which only seemed to produce parties with little to differentiate them, and of the failures of successive coalition governments to resuscitate the economy. As so often in the years prior to the LDP government's collapse in 1993, the Liberal Democrats seemed the lesser of the evils on offer.

Since, with 110 seats, it was still well short of a majority in the Upper House, which will not have elections again until mid-1998, the LDP needed to devise some form of working arrangement with either its former, but numerically decimated, coalition partners or with the new Democratic Party. The DP's leaders seemed unable to make up their minds, and when the LDP turned to its old partners, the SDPJ and the *Sakigake*, both resisted rejoining a coalition. The impasse was finally settled when they agreed to cooperate, on a temporary basis, from outside the cabinet.

The prospect of having to negotiate every piece of legislation with these and other parties, however, seemed to galvanise Hashimoto. He announced an ambitious programme of reform covering five major areas: the structure of government; the economic structure; financial markets; the fiscal system; and the social security system. He then endorsed Finance Minister Hisao Horinouchi's tight-fisted draft budget aimed at raising the consumption tax as a first step towards rebuilding government finances. At the same time, the other political parties have continued to fragment: a leading SDPJ politician left to form yet another splinter group; Hata finally took a clutch of politicians out of the NFP to form the unlikely sounding *Taiyoto* (Sun Party); and the co-leaders of the DP squabbled over whether or not to cooperate with the LDP.

Bureaucracy in the Doldrums

For years the Japanese public has maintained its faith in the honesty and efficiency of the bureaucracy in keeping 'Japan Inc.' running smoothly. Even as a succession of corruption scandals was exposed in the late 1980s and public disillusionment with politicians grew, the Japanese reassured themselves that at least the bureaucrats remained stable. Indeed, they seemed all the more necessary when, after 1993, the LDP was replaced by a succession of coalition governments made up mainly of politicians who had little or no previous experience of ministerial office.

But increasingly in late 1995 and 1996 the Japanese began to have serious doubts about the bureaucracy. Ironically, it is now the politicians who are being called on to begin reform and reorganise the system. Since the Kobe earthquake in January 1995, which revealed flaws in the bureaucratic approaches to approving construction and emergency response, a series of scandals damaged the credibility of these officials. The seemingly all-powerful Finance Ministry has been hardest hit with the exposure not only of its failure to control, but also of its attempt to cover up, malpractices in banking and the securities industry. The police had been severely embarrassed by the nerve-gas attack carried out by the *Aum Shinrikyo* (Supreme Truth) doomsday cult in March 1995 and by the subsequent revelations that it had been a policeman, also a member of the cult, who had then tried to assassinate the country's police chief.

The involvement of Health and Welfare Ministry officials in covering up HIV-tainted blood supplies was exposed and, late in 1996, a top Ministry bureaucrat was forced to resign for accepting financial favours from a company building old people's homes. Senior officials in the Ministry of International Trade and Industry (MITI) were also reprimanded for their links with a tax-evading oil trader. Even in ministries unaffected directly by scandal, bureaucrats admitted to falling morale. Many, therefore, support the idea of some reform, although they invariably tend to believe that reform should start in other ministries rather than their own.

Dancing to the Deregulation Tune

Japan's continued poor economic performance reinforced the calls for reforms in 1996. The government's monthly message of slow but steady recovery was barely believable. A sharp burst of activity in the first quarter of 1996 was followed by another slump, so that gross national product (GNP) growth for the year hovered at around 2%, an acceptable level for most economies but a disappointment to the Japanese whose expectations had been raised by the positive results earlier in the year. Fiscal stimulus packages meant that gross government debt reached 80% of gross domestic product (GDP) by the end of 1995 and was approaching 90% in early 1997. However, net public-sector debt remained low (about 14% in 1996), both in comparison to levels in the 1980s and to those of the Organisation for Economic Cooperation and Development (OECD) member-states. The politically contentious trade surplus did decrease over the year, but still not enough to satisfy Japan's trading partners.

The weakening status of the yen during most of 1996, cost-cutting at home and the increasing shift of manufacturing capacity to cheaper and faster-growing locations elsewhere in Asia did help the larger Japanese multinational companies return to profitability. But the smaller companies, primarily dependent on the domestic market, continued to suffer badly. Polarisation of the economy, with the larger companies becoming stronger and the weak becoming weaker, seemed to be the trend. Unemployment reached an unheard-of 3%. The banks and finance companies were hard hit by the struggle to dispose of suspect loans, and bad debts reached a staggeringly high level.

Nonetheless, despite jitters on the stock market and signs of cold feet from his erstwhile political allies, Hashimoto began 1997 by reiterating his commitment to reform and deregulation as the only way to salvage the economy in the medium term. This is not the first time that Japanese politicians have talked of administrative reform. A decade ago, Prime Minister Nakasone set out on the same path, but made the mistake of loading his policy planning committee with ex-bureaucrats and failed to win over vested business interests. His proposals came to nothing. There are some indications, however, that it may be different this time.

Japan's sluggish economy and persistent bureaucratic bungles have created widespread public support for reform. Big business, on which the LDP still depends for financial support, is far more positive than it was a decade ago. Indeed, the *Keidanren*, the most powerful of the business organisations, has become increasingly blunt in its calls for administrative improvements and the deregulation of key sectors of the economy. Pressure from outside Japan is also a factor; a newly elected and more forthright Clinton told Hashimoto at a meeting in Manila in November 1996 that greater administrative reform and transparency would be a key objective of future bilateral discussions.

The political future of Hashimoto, who has to seek re-election as President of the LDP in September 1997 in order to remain prime minister, depends on accomplishing some successful reforms. His ultimate target is to reduce and realign the functions of the central government ministries and agencies (possibly cutting them down to seven 'mega-ministries'), but it will be several years before such plans can be implemented. Hashimoto needs a higher-profile success in the short term, which is why he has fixed on the idea of a Japanese version of the 'Big Bang' deregulation of the London financial markets. Although Japan's bureaucrats have seen off plenty of ministers in the past who entered office with ideas of change, Hashimoto looks the most likely for some time to carry out real reform.

Foreign Concerns

After several postponements, President Clinton finally made it to Japan in April 1996 and although bilateral economic issues had by no means disappeared, security was at the top of his agenda. The September 1995 Okinawa rape case (for which three US servicemen were sentenced to prison terms in March 1996) had triggered strong anti-US demonstrations on the island, with demands that the Okinawan governor, Masahide Ota, defy Tokyo's instructions to renew leases for the US bases. Some concession by the US became unavoidable, and Clinton agreed that 11 bases and facilities, including Futenma air base, the largest on the island chain, would be returned to Japan within five to seven years.

This agreement, however, was not sufficient to dampen the anti-bases movement in Okinawa. Local elections in June returned a majority of anti-base candidates, and an unprecedented plebiscite in early September also recorded a majority in favour of reducing the US presence even further. Although Tokyo used court rulings to force Ota to sign new leases (which he eventually did), Hashimoto also made a rare prime ministerial visit to Okinawa bearing gifts of substantial financial assistance. In order to maintain the overall size of the US presence while reducing the numbers of troops based on Okinawa itself, the US and Japan devised a plan to build an offshore floating helicopter base (at an estimated cost to Japan of $2bn). The US continues to argue that no actual troop withdrawals will occur, but the momentum of progressive base closures suggests that this will be difficult to achieve without some reductions in manpower.

The Okinawa issue somewhat overshadowed the joint declaration on the continued need for the Japan–US Security Treaty, signed by Hashimoto and Clinton on 17 April 1996 at the mid-point of their summit meeting. Neither side intended to change the Treaty's wording, but the practice of quietly extending what could be done under its rubric continued. The two sides agreed to increase the inter-operability of their forces, particularly in joint operations; to formalise reciprocal provisions of logistical support, supplies and services; to increase exchanges of information and intelligence on the international situation, with particular emphasis on the Asia-Pacific region; and to study ways in which the two countries could cooperate in times of crisis in 'the areas surrounding Japan'. Perhaps somewhat grandiosely, the security partnership was redefined as an Alliance of the Twenty-first Century. Some Asian neighbours, most notably the Chinese and the Koreans, interpreted these measures as indicating an expanded role for Japan's Self-Defense Forces, but the Japanese government argued that it was only carrying out homework it should have done long before.

In 1996, the island extremities of the Japanese archipelago also proved contentious in Japan's relations with other powers. In early February, Japan ratified the UN Convention on the Law of the Sea (UNCLOS), which effectively increased the range of its exclusive economic zone (EEZ) from 12 to 200 nautical miles from its coastline. South Korea was concerned that this would exclude it from fishing around the long-disputed island chain known as Takeshima by Japan and Tokdo by Korea. Korea's response was to build a wharf on one of the islands to aid its fishing fleet, and to mount military manoeuvres in the region. On 9 February in a news conference, Japanese Foreign Minister Yukihiko Ikeda reasserted Japan's claims to the island chain, which South Korean presidential spokesman Yoon Yeo Joon denounced the next day as 'absurd'. Heated feelings on both sides were cooled, however, when Japanese Prime Minister Hashimoto and Korean President Kim Young-Sam, meeting at the ASEM in Bangkok in March 1996, agreed to separate the fishing dispute from the territorial question.

The delicacy of Japan's relations with South Korea was evident in Tokyo's handling of North Korea's request for food aid to help overcome its worsening shortages. In 1995, after careful negotiations with South Korea, Japan pledged 500,000 tons of rice to North Korea. When Pyongyang requested more aid in early 1996, South Korea, which had accused the North of stockpiling the donated rice for its military, was loath to give more without some change in Pyongyang's attitudes towards talks with Seoul on unification. Japan in turn refused the North Korean request on the basis that it had no extra rice available. By early 1997, however, South Korean attitudes had changed again and in mid-February, Tokyo, after consulting with the US and South Korea, was poised to agree to an expected call by the UN for emergency food aid to be sent to North Korea. By staying closely in step with Seoul on this issue, Japan hoped to remove some of the tarnish from its image still produced by its attitudes towards the wartime 'comfort lady' scandal and the territorial dispute over the Takeshima/Tokdo islands.

The Sino-Japanese dispute over the Senkaku/Diaoyu islands proved a more potent issue in 1996. Rather than fishing rights, it was probably the possibility of finding natural resources, such as oil and gas, in the region around the islands that triggered the trouble. The Chinese had been conducting marine research activities around the island for some years, ignoring repeated calls from Japan to desist. While Japan may have ratified UNCLOS partly to enhance its claims to these islands, Tokyo was embarrassed in July 1996 when a small right-wing group, the Japan Youth Federation, built a makeshift lighthouse on one of the islands and planted some memorials and a flag on others. Smouldering Chinese discontent at this action was further fuelled when members of this group were allowed back onto the island on 9 September to repair the lighthouse. This provoked an outpouring of nationalistic sentiment in Hong Kong and Taiwan. Not until one Hong Kong activist had died in the waters off the islands on 25 September and others had succeeded in planting Taiwanese and Chinese flags on one of the islands did the fervour begin to subside. The government in Beijing had tried to contain anti-Japanese demonstrations inside China for fear they might turn into anti-government movements, but left the Japanese government in no doubt that it considered the islands Chinese territory and warned against a resurgence of 'Japanese militarism'.

While the actions of the Youth Federation were not representative of ordinary Japanese people (many of whom hardly cared who owned these islands), the emotional response from Hong Kong and Taiwan was yet another reminder that Japan's relationship with the rest of Asia was still a delicate one. This was driven home in early January 1997 when Hashimoto visited five South-east Asian nations in the hope of paving the way for an enlarged role for Japan in East Asian affairs. The trip to Brunei, Indonesia, Malaysia and Vietnam culminated in a major foreign-policy address in Singapore. In this speech, Hashimoto pushed for a closer partnership with ASEAN countries which would lift Japan's present strong economic ties in the direction of high-level political and security cooperation.

Hashimoto suggested that ASEAN and Japan should institute regular summit meetings, particularly to discuss regional security questions. He also noted that Japan would like to hold 'frank dialogues on regional security with each of the ASEAN countries on a bilateral basis'. Although Hashimoto claimed that each of the leaders he spoke with had agreed to the 'basic idea', it is more likely that his efforts to carve out a new role for

Japan met with polite reserve. The implications of the Japanese initiative are to create a unique position for Tokyo with regard to ASEAN and the rest of Asia. This is certain to offend Chinese sensibilities and to be taken by Beijing as the start of an alliance directed against China. Some ASEAN countries may invite China and South Korea for similar high-level talks to be held in conjunction with the annual meeting of ASEAN top leaders in Malaysia in December 1997, rather than restricting such meetings to Japan.

Is Real Change Likely?

As part of his effort to invigorate Japan's foreign policy, Hashimoto had talks with South Korean President Kim at the end of January 1997, which were also met with reserve. Further fence-mending meetings with Chinese leaders are expected to follow. The difficulty the Prime Minister faces, however, is that most of Asia still feels that Japan does not approach the region with sufficient sensitivity to its suffering during the war. Nor is there any indication that in the near future any Japanese administration, even if it wanted to itself, would be able to gain its population's support for the kind of actions and statements that Asian countries would see as the minimum for accepting a larger Japanese role in regional affairs.

As so often in the past four or five years, domestic problems are likely to figure far ahead of foreign-policy innovations on the Prime Minister's agenda. And it is in domestic economic and social affairs that change is most likely to come. The Japanese are fundamentally a conservative people; they try to avoid the upheavals that change requires, preferring usually to put up with the status quo, even if that is less than perfect. Yet there are times when pressures from below demand change from the top. A ground-swell of that kind is beginning to develop in Japan for economic and political reforms. With the mandarins in the bureaucracy on the defensive, and with increasing support from major business corporations and a majority of the population, Japan's politicians may just find the will to effect the much-needed changes.

Turbulence in the Koreas

Although both North and South Korea had their problems in 1995, it was, on the whole, a relatively quiet year. Hopes that this relative calm might continue during 1996, however, did not last beyond the spring. North Korea's headlong slide into economic decline accelerated, marked by widespread reports of the onset of famine in at least some areas of the country. The political processes in Pyongyang remained opaque. Kim Jong Il assumed none of the positions his father, Kim Il Sung, had left at his death in summer 1994 – except for Supreme Commander of the North Korean military – a number of leadership changes did little to clarify the situation, and the architect of North Korea's governing philosophy defected to the South Korean embassy in Beijing in February 1997 as he returned to Pyongyang from an official visit to Japan. But the deaths of two other senior figures suggested that a generational change was taking place.

Yet there were few signs of a more conciliatory approach to the South. Instead, the North continued its determined effort to deal only with the United States, a position unacceptable to both the US and South Korea. North Korea's continued intransigeance was amply demonstrated when a routine small-scale submarine incursion into Southern waters went disastrously wrong in autumn 1996. The attempted infiltration of armed agents from the North ended in the death of all but one, who was captured, and one who was still at large in mid-March 1997, revealing both Pyongyang's refusal to accept normal relations with the South and the imperfection of South Korea's defences.

The South, for its part, found it no easy matter to come to terms with its past by trying former Presidents Chun Doo Hwan and Roh Tae Woo on corruption and rebellion charges, for the trials inevitably led to questions about the role of the military in both the 1979 coup and the 1980 Kwangju massacre. Other troubles also beset President Kim Young Sam as presidential elections scheduled for December 1997 approached. With both North and South suffering domestic economic and political difficulties there was little sign in 1996 that they would be able, or want, to turn their attention to re-unification; if anything, even though a tentative preliminary meeting finally took place in early 1997, the possibility of meaningful talks on the issue still seemed a considerable way off.

Growing Problems in North Korea

In spite of occasional hints from North Korean diplomats and others supposedly in touch with developments in Pyongyang, Kim Jong Il assumed neither the post of head of state nor that of Party General Secretary in 1996. His only formal positions remained those connected with the military and most of his official appearances and actions have been in a military context. It is generally expected that he will assume the additional titles after the third anniversary of his father's death in July 1997, the end of the respectful three-year mourning period which a dutiful Korean son should honour.

Despite his lack of formal titles, Kim Jong Il still appeared to run the country, and the North Korean media continued to praise him to the full. However, the defection to the South of a senior member of the North Korean hierarchy, Hwang Jang Yop, on 12 February 1997 was seen by many in South Korea and elsewhere as a sign that all was not well in the North Korean leadership. Hwang had been particularly associated with the development of the concept of *juche*, or self-reliance, the North Korean guiding philosophy since the 1960s, and although perhaps only twentieth or so in the leadership hierarchy, his decision not to return to Pyongyang after a visit to Japan was widely seen as evidence of major stresses within North Korea's ruling élite. It was soon clear that this was no sudden decision, further fuelling the belief that all was not well in Pyongyang. Subsequent reports of the retirement of Prime Minister Kang Sung San and the death, from a heart attack, of Defence Minister Choi Kwang – followed quickly by the death of Deputy Defence Minister Kim Kwang Jim – appeared to support the view that Kim Jong Il is grappling with problems at the very top of the leadership. The alternative explanation, that he is ousting the generation that depended on his father in favour of his own protégés, is also plausible. There has been a steady stream of much-publicised defections to the South, but most have been individuals of less importance than was at first apparent, or who had reasons to leave other than a loss of faith in the system. Nevertheless, it is unlikely that the current regime has as full or firm control of the country as did Kim Il Sung.

What is certain is that there has been a steady decline in both the North's general economic situation and its ability to supply food to its population. South Korean sources reported yet another drop in the North's industrial output as the heavy industries that dominate North Korea's economy suffered a shortage of raw materials, a lack of energy, continued distribution problems, and steadily falling productivity as a result of malnutrition among the work-force. These sources estimated the fall in GDP at around 4% per annum since 1991. US and other officials among the small group of foreign businessmen visiting North Korea have now begun to report economic difficulties, with stories of regular power cuts even in the capital, and of factories left idle. Economic planning, once seen as a vital part of the socialist economy, seems to have ground to a halt, and all decisions reflect essentially short-term crisis management.

The North's leaders do sometimes seem aware of the need for new economic policies. Since the mid-1980s, there has been talk of, and even a hesitant move towards, opening up to the outside world. But the pace has been pitifully slow and subject to frequent changes of direction. North Korea is not China, and despite the comments of some in the West, a Chinese-style reform programme based on agriculture would not help the situation. The North has an industrial base similar to those in Eastern Europe, and it requires similar solutions. The regime may realise that it needs foreign capital and foreign know-how, but it seems frightened of the possible political and social consequences of tapping into such resources.

In 1996, the search for additional funds led to negotiations on opening North Korean airspace to foreign carriers for a fee, and to an agreement with Taiwan in January 1997 to store low-grade nuclear waste. This last move was opposed by both South Korea and China, but for different reasons. Other signs of unusual economic enterprise included selling antiques, passing counterfeit US currency and drug smuggling, all of which appeared to have some measure of official support. Continued sales of missiles and other weapons caused international concern.

The regime's main economic hope is the Rajin-Sonbong special economic zone, situated in the far north-east of the country. Many have argued that the zone is so situated to keep the risk of political contamination well away from Pyongyang, but there are also sound practical reasons for its position. It was an area originally developed by the Japanese in the colonial period, and lies close to the Tumen River Development area. If that development should succeed, Rajin-Sonbong would certainly benefit. This is very much in the future and North Korea's main concern has been to attract foreign investment to Rajin-Sonbong in its own right. Given the North's poor record in meeting its foreign debts and the current political uncertainties, this will not be easy, but North Korean officials concerned with foreign trade made a series of overseas visits in 1996 to tout its advantages, and a moderately successful international investment and business forum was held in the zone in September. The most likely investors, South Korean business groups, were banned from attending by the government in Seoul, after attempts by the North to dictate which South Korean officials and journalists could participate.

Increased foreign investment is necessary for the long-term development of North Korea's economy, but of most concern in the short term is lack of food. Despite major efforts to improve crop yields, North Korea has never been able to feed its population without outside resources. Its dependence on these was masked in the past by cheap food supplies from the then socialist countries of Eastern Europe and from the Soviet Union, but the loss of these supplies has left a hole difficult to fill. Heavy rains in 1995 exacerbated North Korea's food problem by extensively damaging the agricultural infrastructure, causing severe food shortages and leading to North Korea's first ever appeals for international aid. Some progress had been made to repair the 1995 damage, although hampered by the population's weakened physical condition, when further flooding in summer 1996 compounded the crisis.

To the structural problems created by unsuitable land and over-concentration on certain crops have been added lack of agricultural technology and fertiliser, and some evidence of the consumption of seed crops thus further reducing the potential harvest. Food rations have either been drastically reduced, even for key groups such as the military, or in some areas apparently stopped altogether. People hitherto used to state-supplied rations have been told to fend for themselves. To help them, there has been some relaxation on former prohibitions on farmers' markets and cross-border barter trade with China, but, as yet, there has been no fundamental change in agricultural production.

Starvation in the North and the possibility of huge movements of people concerned North Korea's neighbours. The Chinese have responded by apparently ignoring their own insistence since the early 1990s that North Korea should pay for assistance, and there were reports of free food supplies to the North. The Japanese hesitated. They have grain available, but found that South Korea was ambivalent about humanitarian aid and, in some cases, exercised a veto over supplies.

The South Korean attitude was mixed. There was concern that fellow Koreans were suffering, and an acceptance that the South should help. But it was more complicated than a simple relief operation. The South had supplied relief in 1995, only to receive little North Korean gratitude in return. Many argued that the North could not expect help while it maintained huge armed forces, threatening South Korea. Others, including President Kim, seemed to take the view that if the North was going to collapse eventually, with South Korea picking up the pieces, it might as well do so now. Seoul was willing to provide food to the North, but only as a result of bilateral talks. The South was still concerned that North Korea could use the threat of famine to drive a wedge between the South and the US, or to establish links with other Western countries and Japan, thus undermining the South's position. In the end, the extent of hunger in the North, and perhaps a reconsideration of what a massive refugee inflow might mean, have overcome Seoul's reluctance to provide aid.

International aid agencies appealed for assistance, but the initial response was less than was needed. Suspicions that North Korea's food shortages were vastly exaggerated were quelled when defectors confirmed the extent of the shortages, and when the International Federation of Red Cross and Red Crescent Societies reported in January 1997 that the official per capita food ration in some parts of the North had been reduced from 200g to 100g per day; 400g is the normal level of humanitarian food supply. That same month, the South Korean Red Cross resumed aid shipments to the North, while in February, the US government announced that it would provide US$10 million through the World Food Programme and South Korea agreed to contribute US$6m to the same programme. Other countries also agreed to help, but the total needed remained much higher than the sums pledged, leaving a real possibility of mass starvation in many parts of the country.

North–South Relations Stagnate

The need for assistance did nothing to modify the North's general approach to North–South relations. The year began inauspiciously, with the North announcing at Panmunjom on 4 April 1996 that it would no longer perform armistice-related duties, followed by intrusions by two groups of armed North Korean soldiers into the Joint Security Area on 5 and 7 April. US and South Korean forces were on higher alert than usual, and South Korean Defence Minister General Kim Dong Jin issued a 'shoot-to-kill' order if there was any attempt to cross the border into the South.

No such attempt took place. The intrusions, probably designed to test Seoul's reactions, were not repeated on the same scale, despite some minor infringements in May along the Military Demarcation Line and North Korean naval vessels intruded into South Korean waters in May and June.

Faced with unremitting hostile propaganda from the North, and a studied refusal to enter into dialogue, the South Korean position became in turn more rigid. Seoul remained concerned that wider US interests in non-proliferation, together with US domestic concern over issues such as the remains of those missing in action from the 1950–53 Korean War, had increased Washington's willingness to deal directly with Pyongyang. Even though US–North Korean negotiations on establishing liaison offices in their respective capitals remained deadlocked, US reassurances that the future of the Korean peninsula could only be settled with the active involvement of the two Koreas seemed to be undermined by direct US contacts with the North on a widening range of issues.

At a summit meeting on 16 April 1996 on Cheju Island in the far south of the peninsula, US President Bill Clinton and South Korean President Kim proposed four-party (China, the US and the two Koreas) talks to the North Koreans. There was nothing to suggest that either China or North Korea had been approached in advance about this specific proposal, although the idea of such talks has been around for some time. While the Chinese agreed that this was one possible way forward, the North Koreans displayed little interest. The Japanese showed some enthusiasm, while Russia demanded that it too be involved. The North Koreans eventually agreed to attend a US–South Korean briefing in New York on 5 March 1997 to explain the concept of the proposal, after postponing the date on a variety of pretexts.

In September 1996, North–South relations took an even more decided turn for the worse when a North Korean submarine, apparently engaged on an infiltration mission, ran aground in Kangwon province on South Korea's east coast. This appeared to be a routine mission of a kind which the North continues to carry out, but one which went very wrong. It provoked real outrage in South Korea, especially when 12 South Korean soldiers, a reservist and four civilians were killed during the subsequent manhunt for the North Koreans who had come ashore. All but two were killed, some by their own side, but the operation had required massive effort and revealed some serious defects in the South's military planning.

Seoul demanded an apology and an undertaking that there would be no repetition of the event. It hinted that if these were not forthcoming, it might seriously consider military action. The North, for its part, refused to apologise. Pyongyang insisted that the submarine had developed mechanical difficulties while on a training exercise, that there had been no infiltration attempt, and that there had been no reason to kill the crew, whose remains should be returned.

The South Korean threat to use force was probably more a mark of frustration than a real indication of a considered change of policy; there have been no signs of the reorganisation of military forces required for such an action, and the conduct of the manhunt showed some clear problems in the South Korean military. Yet the hint was sufficiently worrying for the US to seek assurances from Seoul that no form of independent action was contemplated. President Kim was reluctant to give such an assurance, and at a meeting with President Clinton in Manila in November 1996, he refused to back down on his demand for a full North Korean apology. Mindful of wider concerns, the US pressed North Korea to comply, and on 29 December, for the first time, an apology and an undertaking not to repeat the incursion were issued in Pyongyang. Although the apology was somewhat grudging, it broke the deadlock. The cremated remains of the crew were returned to Panmunjom a few days later.

At various times during the tense negotiations following the submarine incident, the North hinted that the nuclear deal with the United States reached in November 1994 might be off. For its part, the Korean Peninsula Energy Development Organisation (KEDO), created in March 1995 as a result of the deal, began to question the agreement's value in the face of continued North Korean hostility. The submarine incident occurred just as construction of South Korean reactors was about to begin, and brought the project to a halt. Despite such strains, the agreement held, and once the apology was out of the way, KEDO was back on track in 1997. Problems remained over funding, however, especially as costs had increased well beyond earlier estimates, although new members joined the agreement, including the European Union. The other aspect of the nuclear accord – the continued freeze on the North's original nuclear programme – also held; after a pause, the canning of spent fuel rods resumed in January 1997. All these developments indicated North Korea's belief in the growing importance of the United States to its survival. By February 1997, the North was discussing with Washington how to conduct relations at Panmunjom to avoid clashes. One agreement reached was that the US and North Korea would use Panmunjon for sending and receiving US diplomatic mail after they opened their respective offices.

There were few other developments in North Korea's external relations in 1996. Tentative steps to reopen talks with Japan on normalising relations foundered over renewed allegations that the North Koreans had kidnapped a Japanese girl in the 1970s. The Japanese also linked further food aid to this issue. Russia hinted that it wanted to improve relations with North Korea and new friendship agreements between the two countries are expected to be signed in 1997. Pyongyang also held talks with a number of European countries in 1996, but with few concrete results.

China remained the main unknown factor in the affairs of the peninsula. Beijing's relations with South Korea seemed to increase steadily during 1996. Trade and investment links strengthened remarkably, with trade reaching over $20bn, and there were regular ministerial and various other levels of exchanges. The Chinese indicated their doubts about the long-term viability of the North Korean regime. Yet, at the same time, there were hints that the Chinese, like the Russians, saw value in North Korea's continued existence. There were persistent reports that China was continuing to supply grain to the North, and would be reluctant to see a unified Korea on South Korean terms.

There were also indications that China was less than pleased with the activities of South Korean intelligence officers in the Yanbian Korean autonomous area in north-east China. The defection of Hwang Jang Yop in Beijing highlighted China's dilemma. Instead of handling the issue quietly, the South Koreans immediately made it public, perhaps in an attempt to force China to allow Hwang to go to Seoul. China thus found itself in the highly embarrassing position of having to prevent possible clashes between North and South Koreans in their capital, urging the need for calm discussions to solve the issue. All the signs were that China was not pleased with South Korea's behaviour.

South Korea: A Faltering Touch?

Quite apart from the uncertainties over the situation in the North, 1996 was not a good year for President Kim Young Sam, and there were few bright moments. South Korea joined the OECD and, on a different note, the international Football Association agreed to Japan and South Korea jointly hosting the 2002 World Cup. Given the mutual suspicions of Koreans and Japanese, this imaginative solution may cause new problems to be solved by both sides.

For the most part, however, there was a general feeling that the leader had lost his touch. Problems began in April 1996 when the ruling New Korea Party – renamed from the Democratic Liberal Party (DLP) in December 1995 – failed to win an overall majority in the National Assembly elections held on 11 April 1996. While this was not unexpected, and was soon remedied as a scattering of promises brought defections from other parties, it was not an auspicious start. The trials of former Presidents Chun and Roh resulted in the death sentence for Chun and 22 years in jail for Roh. On appeal, these were reduced to life and 17 years respectively, with suspended sentences for various businessmen also involved in the corruption trials.

While many South Koreans welcomed the humbling of arrogant former leaders and the righting of the wrongs of the 1980s, little credit seemed to attach to President Kim. Most were bored by the end of the trials, concerned by an economic downturn and aware that corruption was by no means at an end. Before the end of 1996, a former Minister of Health and a former Minister of Defence had been jailed for corruption, and early in 1997 a further major scandal erupted over the collapse of the Hanbo Iron and Steel Company. Before long, another minister, a presidential aide, a number of parliamentarians and various businessmen had been arrested, and even one of Kim Young Sam's sons was investigated on corruption charges. Although no charges were levied, the scandal gave the opposition an opportunity to allege that there had been a cover-up.

These were not the only difficulties President Kim faced during the year. Violent student demonstrations in Seoul in August 1996 led to the death of one policeman and to claims that pro-North Korean radicals were active among student groups. These claims played on public fears of the North, and the tough suppression measures found favour with the population at large. As a result, Kim Young Sam took a generally more conservative stand on a number of issues. One was the need to strengthen the powers of the Agency for National Security Planning, the former Korean Central Intelligence Agency, to allow it once more to investigate domestic subversion and spying activities for North Korea. Another was a claimed need to reform labour laws. The means to achieve both marked a return to the methods of the past. The National Assembly met at six in the morning on 26 December 1996 with only ruling party members present, and pushed through two bills on labour and national security reform. Although many international observers felt that the labour reforms were long overdue, the method by which the bills were passed outraged the opposition and the unions. Strikes and large-scale demonstrations broke out, which were controlled with difficulty by a show of police force and a promise to reconsider the legislation. Kim's final problem is the December 1997 presidential election and his reluctance to name a successor, possibly for fear of creating an alternative power centre.

An Unsettled Outlook

North Korea is undoubtedly suffering as the cumulative effects of years of economic decline, intensive farming and bad weather take their toll. While few in the international community wish to see people suffer for the follies of their rulers, it has been difficult to drum up adequate support for such a secretive and hostile country. It may be that 1997 will see a return to a more orthodox system of government in the North, although simply the assumption of titles by Kim Jong Il will not in itself bring much improvement. The grave challenges facing the North may force it to accept considerable change both in how it deals with its economic needs and with its relations with the outside world, including South Korea. In the absence of speedy or fundamental economic change, the North seems likely to continue to look to China and the United States to bail it out; if they are not prepared to do so, real chaos may ensue.

Barring the regime's collapse, the increasing links between the United States and North Korea may well keep the nuclear deal, and KEDO, on course, but there are likely to be further moments of high drama as the North tries to win yet another concession for carrying out its agreement and continuing to prepare for the four-party negotiations. The more hardline voices now prevailing in the South will question such concessions, and as South Korea increases the economic gap with the North and feels that it can manage without US support, there is a real risk that Seoul's exasperation with the North's intransigence, or the wish to make a political impact, might lead it to take dangerous chances. Yet the North is still strong enough militarily to make any forceful move very costly, and the consequences for the countries around the Korean peninsula if a conflict should break out are likely to be considerable. Equally worrying is the fear that the North, believing itself pushed into a hopeless corner, might attempt to gain a better bargaining position by military coercion.

For South Korea, a more immediate hurdle is the December 1997 presidential election and the successor to President Kim. He is reluctant to name his successor, as is the custom in Korea, and, after his own treatment of former Presidents Chun and Roh must be well aware of the need to pick a candidate who is likely to win and unlikely to turn against his mentor. There is no guarantee that he can find either.

All in all, the prospects for the Korean peninsula in 1997 look more opaque, and more serious, than they have ever been. North Korea is such a tightly controlled and sealed society that it is difficult for those outside to have any clear idea of which direction it might take. With an election in South Korea, and with growing North Korean suspicions of the South, the United States will find itself drawn more and more into the role of arbitrator between the two Koreas, and may also look for greater international support, perhaps from Europe and Japan, in this new role. What does seem certain is that outside powers will need to assume a greater burden if current tensions are to be controlled.

The Weakening Consensus in the Asia-Pacific

The shape of ASEAN and its policies has changed in the past year. The Association has now pledged to expand to include all Indochinese states and Myanmar by 2000. This larger ASEAN will seek to retain its leading diplomatic role in the ARF in which all the major Asian powers, including India since 1996, participate. Recognition of ASEAN's continuing relevance came with Japanese Prime Minister Ryutaro Hashimoto's suggestion on a visit to South-east Asia in January 1997 of annual summit meetings with Japan. Yet, as was made clear by the security agreement which Indonesia and Australia concluded in December 1995, some ASEAN states are beginning to flirt with bilateralism. ASEAN pledged to expand and apparently prepared to locate much of its diplomacy in a strengthening

ARF, may well struggle to maintain the intimacy, and therefore coherence of decision-making, that characterised its earlier days.

ASEAN was spurred to reassert its separate regional identity in part by the relative success demonstrated by the Asia Pacific Economic Cooperation (APEC) forum – a multilateral group committed to regional free trade and investment. Since 1993, at US initiative, APEC has combined annual ministerial meetings with meetings of heads of government (with the exception of Taiwan and Hong Kong which have a diminished status within the group). Even discounting the degree of economic cooperation that has so far been achieved, APEC has provided an important forum for bilateral security dialogues at the highest levels. For example, at its summit in Subic Bay, Philippines, in November 1996, both US President Bill Clinton and President Fidel Ramos of the Philippines held private discussions with Chinese President Jiang Zemin. In November 1996 in Jakarta, ASEAN also established annual informal meetings of heads of government, changing its club of foreign ministers into a grouping in which the member-states' major political decision-makers assumed more direct charge of the Association.

An Enlarged ASEAN Regional Forum

The ARF convened for its third working session in Jakarta in July 1996 with no substantive advances in its modest programme of confidence-building through inter-sessional dialogues. Agreement was reached to continue the Inter-Sessional Support Group on Confidence-Building Measures for a further year, and China offered to co-chair with the Philippines the Group's next meeting in Beijing in March 1997. The Inter-Sessional Meetings on Search and Rescue (co-chaired by Singapore and the United States) and on Peacekeeping Operations (co-chaired by Malaysia and Canada) were endorsed, and provisions were made for another on disaster relief to be co-chaired by Thailand and New Zealand. It was also agreed to convene two 'track two' meetings (ostensibly among non-officials) in Paris and Jakarta to discuss preventive diplomacy and non-proliferation. The Chairman's concluding statement alluded to differences between ASEAN and China over the South China Sea, and noted that 'there was some divergence of views on the subjects discussed', but overall its tone was positive, claiming that the participants had 'displayed a high degree of comfort in their interactions with each other'. The ASEAN states also gained the general endorsement of the South-East Asian Nuclear Weapons Free Zone (SEANWFZ) Treaty which had been concluded in Bangkok in December 1995 despite criticism from China and the US.

Although there was some plain speaking over the South China Sea and the CTBT, the main argument in Jakarta centred on the procedure followed to admit India into the ARF, and the entry of Myanmar into ASEAN.

Although separate, both issues highlighted a conflict of interests between ASEAN and both Asian and non-Asian participants in the ARF. It was strongly felt by non-ASEAN members that the Association had been presumptuous in presenting a virtual *fait accompli* concerning Indian membership to a May 1996 meeting of ARF senior officials in Indonesia. Bringing India into the organisation has considerably extended its geographic scope and threatens it with new and contentious issues. The plan to include Myanmar was controversial for different reasons, but it too has upset many in the ARF as discussed below.

At a meeting of ASEAN heads of government in Bangkok in December 1995, India had been given the status of 'dialogue partner'. This term was a relic of the late 1970s when industrialised states held regular meetings with an ASEAN interlocutor as well as attending so-called Post Ministerial Conferences after the annual meetings of foreign ministers. At this December meeting, Singapore, supported by Indonesia, pressed India's case in part to counterbalance China, especially in relation to Myanmar which some feared was falling into a Chinese sphere of influence and which had become a candidate member of ASEAN in July 1995 on signing its Treaty of Amity and Cooperation. All ASEAN dialogue partners had automatically become participants in the ARF when the Forum was established in 1994. ASEAN felt that India had a strong case for participation, and believed that New Delhi would support the Association's diplomatic centrality within the ARF, which had become a matter of some dispute. There was considerable resentment, particularly in Washington and Tokyo, that ASEAN had taken its fellow ARF members for granted and had failed to display normal diplomatic courtesies in promoting India's participation – possibly in an attempt to stress that the Association was still 'the prime driving force' within the Forum.

In the event, India's admission to the ARF became a formality at the Jakarta working session. It was justified in terms of its location within the 'geographical footprint' of key Forum activities, and because it could 'directly affect the peace and security of the region'. However, the process by which India entered the ARF left a sour taste that drew attention to the need for consultation and consensus over the procedure for admitting new participants. At stake was the extent to which the major Asia-Pacific powers would allow ASEAN to drive the ARF. But the major powers themselves could not agree, and this worked in ASEAN's favour. While the United States and Japan wished to limit the Association's diplomatic centrality within the Forum, China and India saw the political utility of conciliating ASEAN, thereby preventing the US from dominating the ARF's agenda and activities. Indeed, the absence of a concert of major powers acting in a managerial capacity within the ARF is the key to ASEAN's enhanced status and role within the multilateral dialogue.

An Enlarged ASEAN

The commitment ASEAN made in Bangkok to bring all ten South-east Asian states into the organisation by 2000 was partly based on its desire to realise the vision of all-inclusiveness set out in its founding declaration in August 1967. Underlying that commitment, however, was an interest in demonstrating a united South-east Asian front within the ARF and, in the case of Myanmar, to provide a mechanism to counter Beijing's influence. China, which concluded a new agreement with Yangon at the end of 1996 to increase military cooperation – including air force and naval training and arms sales – was a major concern. The admission of Laos and Cambodia to ASEAN was not controversial. Myanmar, however, with its notoriously poor human-rights record was not an acceptable partner to the ARF's Western members. ASEAN had adopted a policy of so-called 'constructive engagement' towards Myanmar on the grounds that isolating the government in Yangon, especially through economic sanctions, would serve no practical political purpose. In late May, the regime in Yangon had made mass arrests of members of the opposition National League for Democracy led by Nobel Peace Prize laureate Aung San Suu Kyi. In June, James Nichols, an Anglo-Burmese businessman and close associate of Aung San Suu Kyi, who had been serving three years in prison for using unauthorised fax and telephone lines, died while in custody. Nichols had acted as an honorary consul for Norway as well as representing Danish, Finnish and Swiss interests. His death had a significant international impact, precipitating concerted diplomatic action by the European Union against Myanmar at the ARF, supported by Australia, Canada, New Zealand and the United States. Matters came to a head at the ARF in Jakarta in July 1996. Although nothing could be done about Myanmar's participation in the Forum, an attempt was made to penalise the state by urging ASEAN not to admit it to full membership, even though it had just been accorded observer status.

Myanmar formally applied to join ASEAN in August 1996 during a visit by Prime Minister Than Shwe to Malaysia which, as Chair of ASEAN's Standing Committee for 1996–97, had a special interest in seeing the vision of a ten-member ASEAN fulfilled at the 30th anniversary conference of foreign ministers to be held in Kuala Lumpur in December 1997. That formal request prompted second thoughts within the Association about the timetable for admission. ASEAN's heads of government held their first annual informal summit in Jakarta at the end of November 1996 and Prime Minister Than Shwe attended as an observer. The participants reiterated their decision to admit the country despite Western attempts to influence their agenda, but pointedly avoided setting a specific date. It was evident, however, that if Myanmar were not admitted to ASEAN at the same time as Laos and Cambodia, it would be seen as being isolated diplomatically against the spirit of 'constructive engagement'. In an attempt to avoid this problem, the heads of government agreed that Myanmar, Cambodia and Laos would all be admitted 'simultaneously', in the hope that the government in Yangon would be sufficiently accommodating in its domestic politics to permit all three to enter in July 1997. In January 1997, reservations held by the Philippines and Thailand about Myanmar's membership were withdrawn and Malaysian Prime Minister Dr Mahathir Mohamed confidently predicted Myanmar's entry into ASEAN during the coming year.

Enlarging ASEAN to include Myanmar has caused the Association severe political embarrassment. While Myanmar's ruling State Law and Order Restoration Council (SLORC) has broken the back of long-standing ethnic rebellion and upheld unity within its military junta, it has been unable to reduce the moral authority of Aung San Suu Kyi. Yet, widespread student demonstrations at the end of 1996, together with two bomb explosions at a Buddhist shrine in Yangon on Christmas Day, had no visible impact on the political staying power of the SLORC. By contrast, Vietnam, which joined ASEAN in July 1995, has adapted well to the Association's diplomatic culture. Although tensions between Vietnam and China still exist, the bilateral relationship has settled into one of working accommodation exemplified by the visit to Hanoi in October 1996 of a delegation from the International Liaison Department of the Communist Party of China. The Eighth National Congress of Vietnam's Communist Party in June–July 1996 highlighted the difficulty of reconciling market-based economic reforms with a conservative political system that failed to make any critical changes in a gerontocratic party and national leadership. Laos faces a similar dilemma to Vietnam over its social and economic system, and its ageing leadership, but like its eastern neighbour it continues to be politically stable.

Security is an acute issue only regarding the third candidate, Cambodia. Cambodia is torn by a rift within the coalition government formed in the wake of UN-supervised elections in 1993. Although the Cambodian People's Party (CPP), imposed initially by Vietnamese force of arms as the Kampuchean People's Revolutionary Party, secured fewer seats than its royalist rival *Funcinpec*, it has retained effective control of security and administration, especially at the provincial level, despite promising to share power. The relationship between First Prime Minister Prince Norodom Ranariddh of *Funcinpec* and Second Prime Minister Hun Sen of the CPP has gone from bad to worse. Both sides competed for the main role when a total of about 10,000 dissident elements of the Khmer Rouge under deputy leader Ieng Sary defected in August 1996 (see map, p. 248). In January 1997, Prince Ranariddh announced that he had created a new United National Front comprising *Funcinpec*, the Buddhist Liberal Democratic Party (a dissident faction of a minor coalition partner), the unregistered Khmer Nation Party and a defector wing of the Khmer

Rouge, thus creating even sharper polarisation in Cambodian politics. The same month, Hun Sen told an extraordinary congress of the CPP, also attended by defectors from the Khmer Rouge, that he had rejected Ranariddh's offer to join the new front. As political competition intensifies prior to elections in 1998, internal conflict may provoke renewed external intervention.

Indonesia's Predicament

ASEAN's security doctrine and practice rejects defence cooperation under its aegis, and Indonesia was a major influence in adopting this doctrine. It is a founder member of the Non-Aligned Movement and was its chair in 1992–95. It describes its foreign policy as 'independent and active', although Jakarta has not concealed its interest in obtaining a permanent seat on the United Nations Security Council. Indonesia also strongly believes that the internal affairs of member-states should not be the subject of international attention and 'meddling'. For that reason, it has supported Myanmar's membership of ASEAN driven, in part, by its own difficulty in coping with the absorption of East Timor.

Indonesia was acutely embarrassed by the joint award of the Nobel Peace Prize in October 1996 to Bishop Carlos Bello, resident in Dili, and *Fretilin* resistance leader José Ramos-Horta, in exile in Australia, for their challenges to its occupation. The issue of East Timor has led to tensions in Indonesia's and ASEAN's relations with the European Union – a member of the ARF – because Portugal is pressing the issue. It has also not helped relations with the United States. Revelations of Indonesian funding for the Democratic Party's presidential campaign, and of personal links between an Indonesian financier and President Clinton which allegedly influenced US policies in Asia, have aroused Congressional interest.

Indonesia has been consistent, with other ASEAN states, in fending off international intrusions into its domestic affairs. In security policy, however, President Suharto has begun to indicate a new appreciation of the balance of power, anathema to conventional wisdom within the Foreign Ministry. The security agreement concluded with Australia in December 1995 contained *inter alia* the language of an alliance in direct contradiction to Indonesia's official foreign policy. That agreement, negotiated in utmost secrecy and coming as it did after instances of Chinese maritime assertiveness in the South China Sea, could hardly be divorced from them. In addition to the revelation in February 1995 of its occupation of Mischief Reef in the Spratly Islands, China appeared to have laid claim to maritime jurisdiction within Indonesia's exclusive economic zone arising from the Republic's possession of the Natuna Islands at the western periphery of the South China Sea (see maps, pp. 246–47).

The dark shadow that a rising China is casting over the region has been an important factor in Indonesia's change of attitude towards the policy of cooperative security. In September 1996, Jakarta attempted to make clear to Beijing that it firmly intended to protect its maritime and resource interests by conducting a large-scale two-week military exercise on and around the Natuna Islands. It was hardly a coincidence that at the same time Indonesian Foreign Minister Ali Alatas received his Taiwanese counterpart, John Chang, even though he claimed that the meeting had no official status. The previous month, President Suharto had taken the unusual step of commenting publicly on the prospect that China might become a threat to Asia, although he couched his concerns in economic terms of the danger of low-priced goods flooding regional markets.

Although Indonesia has been a committed member of ASEAN since its inception, its role within the ARF has been ambivalent because of its concern that the new security dialogue might act as a Trojan horse for the intervention of major external powers. China is Indonesia's prime worry in the medium term, which explains its engagement in activities that complement cooperative security, including its burgeoning security relationship with Australia. Australia itself entered into a strategic partnership with Singapore in January 1996 and increased its security cooperation with the United States in an agreement concluded the following July. Underlying the change in Indonesia's security policy is an acceptance of the need for a US military role in the Asia-Pacific to counter China's likely assertiveness as well as any change in Japan's long-standing security policy. This position is shared to differing degrees within ASEAN, although Thailand and Malaysia are more confident than other members of being able to manage bilateral relations with a rising China.

Increasing domestic political unrest has also shaken Indonesian confidence. Rioting broke out in Jakarta in July 1996 and had to be quelled by a massive show of police force. In retrospect, this may have been a contrived attempt by the government to crush a rising opposition movement led by Megawati Sukarnoputri, daughter of the late President Sukarno. Part of the problem is the undercurrent of uncertainty created by President Suharto – who turned 75 in June and whose wife died the previous April – who has given every indication of seeking a further five-year term of office in elections scheduled for March 1998. Social unrest arising from economic grievance was also expressed in riots elsewhere in Java in October and December 1996 directed against Christians, local Chinese and notably the police, and in ethnic violence in West Kalimantan between indigenous Dyaks and migrants from the island of Madura in January 1997. Further violence occurred in two areas of Java and in the West Kalimantan capital, Pontianak, in January 1997. With the prospect of renewed disturbances coinciding with parliamentary elections in May 1997, how the ultimate resolution of the political succession might affect Indonesia's approach to regional security cooperation – very much the personal policy of President Suharto – is of major concern.

One reason for Indonesia's change in its foreign-policy philosophy is that it finds ASEAN too small a vehicle for its ideal international role, and the ARF, given the presence within it of the major Asia-Pacific powers, too large. Nonetheless, Indonesia remains ASEAN's effective anchor. If it were to come adrift politically because of intra-Association instability, the repercussions for regional security would quickly manifest themselves.

The Prospect

Although signs of differences have arisen concerning how best to effect cooperative security in Asia–Pacific, a novel multilateral dialogue has been sustained, encouraged by strong economic interdependencies. The present sense of a secure area is reinforced by the fact that the domestic situation within regional states offers little opportunity for external competitive intervention, while the post-Cold War external environment provides no ready context in which to revive that dangerous prospect. To be sure, critical issues such as the status of Taiwan and the Korean peninsula remain outside the ambit of cooperative security. Nor does the ASEAN practice of cooperative security provide a direct mechanism for resolving disputes in the South China Sea or maritime issues between Japan and China and Japan and South Korea. Yet the fact that the countries of the region have made the ARF's meetings a matter of routine is symptomatic of the measure of regional consensus underlying the process of dialogue. Even the limits of cooperative security, created by its very nature, are only too well understood, and it was this understanding which led to the Indonesian–Australian bilateral agreement. Unless China can be induced to moderate its behaviour in line with its willingness to engage in a dialogue about consensus, a number of other nations in Asia-Pacific may follow this model of balance-of-power politics.

South Asia: Looking Inward

National-security issues were of secondary concern on the subcontinent in 1996, as election politics, government-building and economic developments dominated attention in India and Pakistan. The April–May general election in India was strongly influenced by a widespread corruption scandal and, although once elected, the new government showed some strength in its formulation of foreign and economic policy, it governed largely at the sufferance of the Congress Party. The commitment of the Congress Party and its previous government to economic liberalisation helped buttress India's continued economic expansion, but this did not translate into electoral support. It has, however, led to the Congress Party's full support of the coalition government's continuation and even expansion of economic reforms.

Pakistan's election in February 1997 was conducted against a background of continuing domestic violence and fiscal mismanagement. After President Farooq Leghari deposed former Prime Minister Benazir Bhutto in November 1996 he appointed an interim government. This government, however, failed to fulfil its promises to hold all corrupt officials accountable, the ostensible reason for relieving the Prime Minister in the first place. Although the new government of Nawaz Sharif emerged from the election with a large parliamentary majority, it may not prove much more capable of undertaking vital reforms than India's government. Sharif won with a very low voter turnout, and must adjust to working with the newly formed Council for Defence and National Security (CDNS). Economically, Pakistan continued to be saddled with debt servicing and spending which consumed in real terms at least 60–70% of the budget. In neither India nor Pakistan did elections bring much hope of a short-term solution to their underlying problems.

Finding Leadership in India

National elections, held from April to May 1996, confirmed that without a powerful Nehru/Gandhi family member leading it, the Congress Party has run into trouble. Prime Minister P. V. Narasimha Rao was defeated mainly by the pervading corruption that surrounded his office, but a number of other factors also motivated the opposition. To his credit, Rao had steered India towards greater involvement in the international economy. But his ability to translate an international economic vision into domestic political success was notably lacking. That he succeeded in liberalising the economy did not help him. If anything, it underlined more than ever the differences between India's more and less developed areas, and the Congress Party paid a political price for appearing to favour the rich over the poor. At the same time, Rao demonstrated lamentable political judgement. One of his most questionable decisions was to back the ruling party in Tamil Nadu, whose leader Jayaram Jayalalitha was subsequently arrested on tax-evasion charges. As a result, Rao not only lost the state's votes, but also forfeited the support of Palaniappan Chidambaram, his former Finance Minister, who subsequently assumed the position of Minister of Finance and Company Affairs in the new government. Rao was also under scrutiny by the Supreme Court for involvement in the corruption scandal. It was this combination of events that contributed to the turn of popular sentiment against the Congress Party.

After the election, Congress had too few seats to form a new government, although the 140 seats it managed to win was enough to ensure a central role in any coalition. Sitting ideologically between the conservative, Hindu-nationalist *Bharatiya Janata* Party (BJP) and the National Front–Left Front, Congress was in a position to give its support to whichever party took power. Without Congress' support, no government would last.

The toppling of Congress in the elections was heavily motivated by protest politics, as the electorate responded with disgust to the widespread corruption. Underlying its defeat, however, was some questioning of India's national identity, as the BJP emerged with the most seats at 161, despite the fact that its leader, Lal Krishna Advani, had also been implicated with key Congress leaders in the Jain/*hawala* money laundering scandal. Given its commitment to a Hindu India and *Hindutva*, or Hindu nationalism, the BJP poses a challenge both to the Congress Party's secularism, and to the larger vision that Congress has nurtured since India's independence 50 years ago. But just as Rao has been unable to translate an economic vision into a political mandate for the country, so too the BJP has been unable to turn its vision of a Hindu India into a national political mandate. It has expanded from its base in Maharashtra, and scored impressive gains in Uttar Pradesh and Bihar, but it still lacks support across the breadth of India.

The controversy that surrounded the BJP's vision was made evident by the opposition it aroused rather than the support – only 21% nationwide – it gained. Nevertheless, since it won the most seats it was asked to form a government. But opposition to a BJP government was so pronounced that, after two weeks of seeking coalition partners, it chose to resign rather than face a no-confidence vote in the Lok Sabha. That such a vote was inevitable had been made clear even before Indian President Shankar Dayal Sharma asked the BJP to form a governing coalition, as during the election campaign Congress had declared that it would not support a BJP government. As the party with the largest vote, slim though its plurality was, the BJP was given a chance which in the event proved more illusory than real. Congress made good on its threat, as it could not support a party which so publicly contradicted what it had stood for for so long. Congress had made deals in the past, but this was one it had to resist.

The ultimate winner – almost by default – was the National Front–Left Front coalition, a loose affiliation of 13 regional parties with little more in common than an aversion to both the BJP and Congress. Indeed, once the election was over, the coalition was referred to as the 'third force' until it adopted the name United Front (UF) to represent its united opposition to the other two parties. Finding a parliamentary leader among a group of regional party bosses proved difficult, however. Former Prime Minister V. P. Singh refused to assume the position again, due to failing health. Surprisingly, the coalition then turned to the octogenarian Bengali Jyoti Basu, leader of the West Bengal-based Communist Party of India (Marxist) (CPI(M)). The problems of coalition politics were swiftly made evident as the CPI(M) party leaders refused to support Basu as Prime Minister because they feared being tainted by compromise. It then fell to H. D. Deve Gowda, a local politician from Karnataka, to take up the reins. The fact that Deve Gowda spoke no Hindi, while his new Defence Minister and

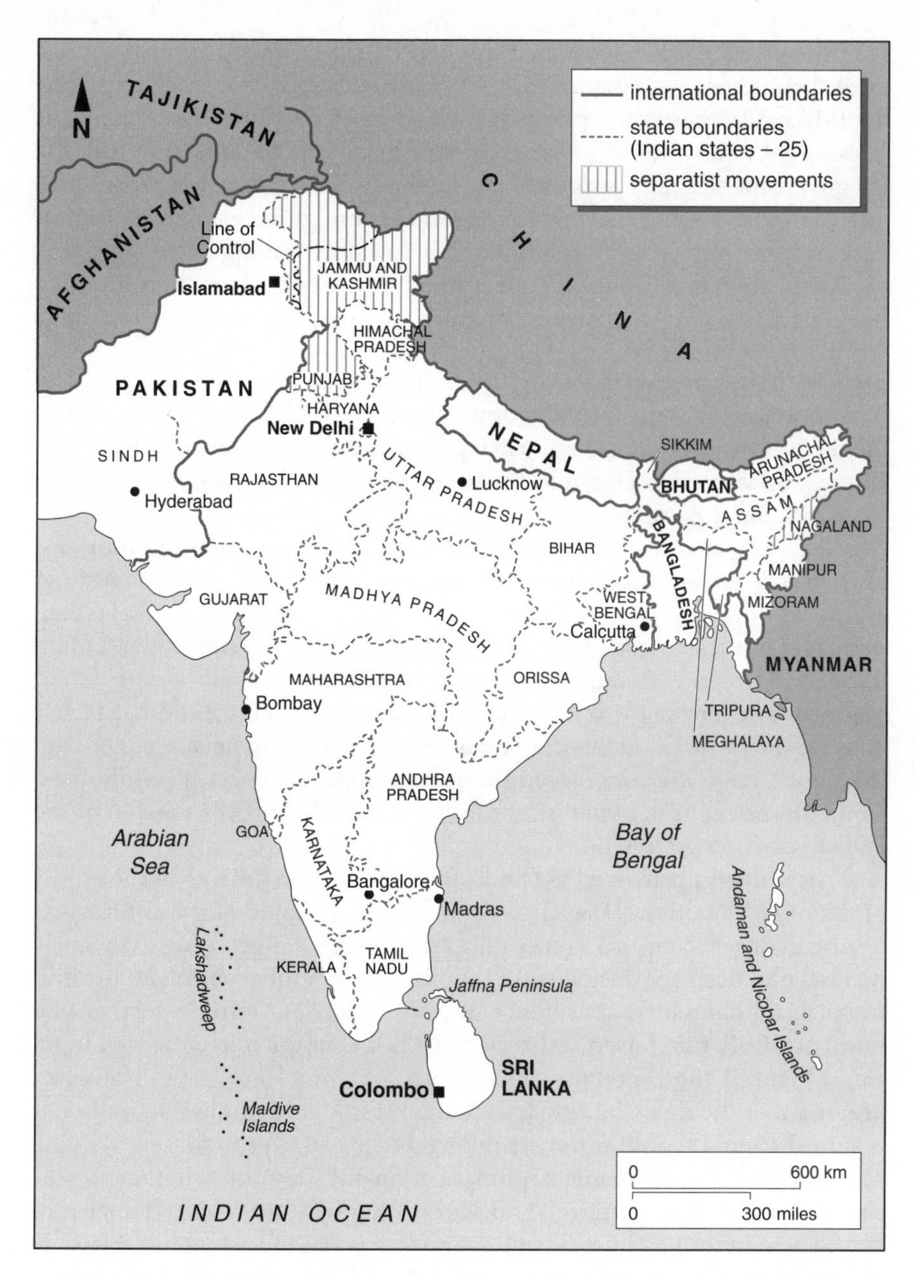

coalition partner, the former wrestler and Uttar Pradesh party boss Mulayam Singh Yadav, spoke no English, made government deliberations all the more intriguing.

The unlikely UF has, however, managed to coalesce and form a government. It has passed relatively little new legislation, but has made a clear commitment to the economic liberalisation programme that its predecessor put into effect. The economy continues to do well, recording a 6.6% GDP growth rate for the first half of fiscal year 1996–97, following the record-setting 7% for 1995–96. Support for state-funded industries still poses problems, but Finance Minister Chidambaram in discussions with the IMF in October 1996 optimistically promised that fiscal deficits would be brought below 5%. Another economic bright spot was that in July the UF government finally approved the landmark Enron/Dhabol power project, perhaps clearing the way for expanded foreign investment in India's surging economy. This billion-dollar US investment in India's energy sector had been delayed since the BJP gained power in Maharashtra in 1995.

National-Security Issues

Security issues on two fronts occupied India's attention during 1996. The most public was India's role in opposing the CTBT, but the more fundamental issue may have been the marked improvement in its relations with China, punctuated in December 1996 by the visit to New Delhi of Chinese President Jiang Zemin.

The year began, however, with a kind of intellectual 'mopping up' after India's late-1995 flirtation with resuming its nuclear-weapon testing. At the end of 1995, India was detected preparing its site in the Rajasthan desert for a nuclear test. Once the media-created stir calmed down, Indian Foreign Secretary Pranab Mukherjee admitted that the country had planned 'to exercise the option', but, he added, something had happened and 'we may exercise it later'. In other words, a test had been planned, but the public response had forced the government to cancel it. While New Delhi did not conduct the test, it did test-fire the Air Force version of the *Prithvi* short-range missile.

These events provoked yet another debate within India over the merits of the nuclear option. Hawks contended that it should carry out nuclear testing in order to ensure a robust nuclear deterrent. According to them, as no tests had been conducted since the single 'peaceful experiment' in 1974, India could have little confidence in its nuclear deterrent posture until a more carefully diagnosed test series had been completed. Others in India clearly feared the repercussions of so open a nuclear stance, however, arguing that the financial restrictions the West would impose on India if it resumed testing would seriously retard its economic growth.

Whether the economic argument won the day, or whether it was simply decided that additional nuclear testing was not required to ensure confidence in India's nuclear option, no test was conducted and the issue was replaced by wrangling over the CTBT in Geneva and New York. This debate again highlighted a number of important issues for India: whether its single test in 1974 was a sufficient basis for nuclear deterrence; whether India should – or could – develop boosted or thermonuclear weapons; and what the trade-offs were between economic policies designed to improve India's domestic security and well-being, and security policies designed to keep China and Pakistan at bay.

At the end of 1995, India had been disappointed at the indefinite renewal of the Nuclear Non-Proliferation Treaty (NPT), although this was hardly unexpected. The more hawkish response from New Delhi had been to argue that nuclear weapons in the hands of the five declared nuclear-weapon states (NWS) were thereby made permanent – in effect, that the NPT had legitimised nuclear weapons in the hands of the few, rather than committing the NWS to nuclear disarmament under the terms of Article VI. This reasoning was unconvincing to the outside world, but resonated for many within India, propelling it into its next major international confrontation, the CTBT negotiations at the Conference on Disarmament (CD). While India had been a non-party to the NPT talks, as a Conference member it was in a better position to make its views on the CTBT heard.

Once it was clear that the five NWS supported a CTBT – and once the French and Chinese in late spring 1996 had completed their last rounds of nuclear testing – debate in Geneva turned on whether laboratory-scale or sub-critical testing would be allowable under the terms of the Treaty. As with many other treaties, the US position was to avoid language that in effect begged the question by claiming to restrict activities that could not be detected and properly verified. Allowing such tests, however, according to India and a number of other delegations, would permit continued improvements in the stockpiles of the declared NWS. India furthermore objected to the entry-into-force provisions, which required signatures by the three nuclear-threshold states – Israel, Pakistan and India – and also objected because a time-bound framework for eliminating all nuclear weapons was rejected. India also noted that its neighbours were increasing their nuclear-weapon capabilities, and argued that for security reasons it could not afford to sign the CTBT. New Delhi therefore blocked a consensus on the Treaty text in the committee report to the General Assembly (UNGA).

Faced with this position, Western nations, in an unusual response reflecting their feeling that India had betrayed its historic opposition to nuclear testing, agreed to submit the Treaty to the UNGA directly. There it received enthusiastic support, thus making India's blocking manoeuvre at the CD moot. In the vote, only Bhutan and Libya joined New Delhi in opposition. India's diplomatic defeat could barely have been more complete, and the UNGA resolution bringing the CTBT into force was passed 24 September 1996 despite India's objections. Ironically, India had been one of the first advocates of the idea some 40 years earlier.

India's strong opposition to the CTBT provided Pakistan with substantial cover on this issue. Pakistan has proposed a number of arms-control agreements over the years, perhaps in the knowledge that India would resist them. With New Delhi vigorously opposing the CTBT, Pakistan agreed to sign it – if India would do so as well. Despite the diplomatic embarrassment it suffered in New York, India paid a relatively low cost for its strong stance against the CTBT. The US was preoccupied with its own election for the latter half of the year, and paid little heed to South Asia.

Regarding China, Sino-Indian expert meetings were held in New Delhi and Beijing in 1996 to discuss the border dispute, and a joint working group concluded an agreement in October to defuse concerns along the border. Although New Delhi apparently continued to resist China's proposals that the line of actual control separating the two countries be jointly recognised as a permanent boundary (see map, p. 249), the state visit by President Jiang Zemin in December reinforced the impression that diplomatic engagement was a more productive way of ensuring security relations along the northern border than nuclear weapons ever would be.

India was active on other diplomatic fronts as well. During a visit by Russian Defence Minister Igor Rodionov, New Delhi signed a cooperation agreement with the Russian Ministry of Defence. Looking east as well, India became a formal member of the ARF. Although these diplomatic efforts improved India's security environment, their impact was not enough to convince law-makers to cut defence spending. Instead, the annual budget included a 20% increase over the previous year, although much of this will be spent on pay rises.

Contrasted with the relative amity on the northern and eastern fronts, however, was the continuation of diplomatic silence with Pakistan. The new Indian Foreign Minister, the highly respected Inder Kumar Gujral, took steps to end the brittle stand-off between the two governments. He offered to resume talks at foreign-secretary level, which had broken off in early 1994, but was effectively rebuffed by Prime Minister Bhutto, who said that Gujral's proposal presented 'difficulties' since it did not put Kashmir at the top of the agenda for discussions.

The offer might have been taken up by the interim government that replaced Benazir Bhutto in Islamabad in November, which included Sahabzada Yaqub Khan as Foreign Minister. Yaqub and Gujral had headed their respective diplomatic missions to Moscow in the 1980s, where they came to know and respect each other, and had subsequently each held the position of Foreign Minister in their respective countries (this is a second term for each of them). With elections due in Pakistan, however, foreign relations were put on hold, and the diplomatic freeze continued.

Although not all of India's efforts were successful in 1996 – it tried, but failed, to be made a permanent member of the UN Security Council – by the end of the year, it was moving into position to escape the consequences of its Cold War backing of the Soviet Union. In economic terms, its international role was expanding; in political terms, it had defended an unpopular position but survived to be wooed by the United States; in security terms, it had retained its independent nuclear option against widespread international opposition; and in diplomatic terms, it had strengthened relations with its most powerful neighbour.

The Pakistani Merry-Go-Round

Pakistan's domestic problems worsened considerably during 1996. Violence was endemic and the government seemed to have no viable way of handling it. The economy was in tatters, and once again President Leghari dismissed the government in power on grounds of corruption, called another election, and then swore in a government led by the party that had been dismissed for similar reasons only two years earlier.

The chaotic conditions in the country were epitomised by Karachi, where violence since January 1994 has claimed almost 3,000 lives even by official estimates. In his effort to gain control over the crisis-ridden city, however, Interior Minister Naserullah Babar may have sowed the seeds of deeper future problems. Babar's approach during 1996 was effectively to unleash the Karachi police on the *Mohajir Quami* Movement (MQM), the principal political force in the city now composed predominantly of *Mohajirs*, or Indian immigrants. It soon became clear that the police in their zeal were prepared to shoot first and ask questions later, if at all. Although the short-term result was a significant decline in violence attributable to the MQM, the government did little to address the structural economic and political grievances that had originally fuelled support for the Movement. Babar may have thought that brutal suppression would succeed in Karachi as it had in India when it confronted Sikh terrorists in Punjab. But where the Sikh terrorists had been a small minority in a large, essentially rural Indian state, the MQM dominated a densely populated urban centre. In this case, similar tactics are unlikely to produce similar results.

While the government was carrying out what were widely considered to be 'extra-judicial' killings in Karachi, it was soon faced with additional domestic problems. Conflict between the Sunni and Shi'i increased in number and violence in Punjab, Pakistan's heartland. Bombs exploded both in the Shaukat Khanum hospital in Lahore and at Lahore International Airport, as well as on a bus outside the city, leaving over 50 dead. In addition, there were a number of less spectacular but still deadly incidents. While the violence was confined to Karachi and Sind, it seemed less of a threat to Islamabad. But as violence erupted along internal Islamic divisions, Pakistan's integrity was threatened more profoundly.

The spiral of violence engulfed Prime Minister Bhutto's brother and political rival, Murtaza Bhutto, who was himself gunned down by Karachi police in September 1996 in a late-night confrontation outside his home. The Prime Minister's husband, Asif Ali Zardari, was immediately assumed to be behind Murtaza's death. Astonishingly, no official investigation of the murder was completed before Bhutto herself was sacked by President Leghari in November. Once deposed – but not before – Benazir pointed an accusing finger at the President himself. It was not at all clear how Leghari might benefit from eliminating the last remaining adult Bhutto male, who was a potentially viable opponent of the Prime Minister he later removed from office, and the accusation received little credence.

Bhutto's second tenure as Prime Minister had given Pakistan an opportunity to mature in democratic terms. Former President Ghulum Ishaq Khan had deposed Bhutto's nemesis and predecessor, Nawaz Sharif, in April 1993, much as he had deposed Bhutto after her first term in August 1990. When Sharif challenged his dismissal, the Pakistan Supreme Court ruled in his favour, finding that the dissolution was illegal. It was a banner day for the rule of law. The military then negotiated a face-saving exit from the crisis for Ishaq Kahn and Nawaz Sharif, as both resigned. New elections were called, and Bhutto swept to victory amid a sense that Pakistani democracy had matured.

That euphoria was quelled by Bhutto, however, once her new government was seated. She retained the finance portfolio for herself, and proceeded to mismanage the economy, leading Pakistan into its worst economic position in its history. With her husband Zardari widely perceived to be feathering his own nest at the public's expense, Bhutto audaciously thumbed her nose at the Pakistani community by reshuffling her cabinet in July 1996 and appointing Zardari to the newly created position of Minister of Investment. Given that he was allegedly already extracting bribes as high as 30% for government contracts, the Prime Minister's decision to give him official responsibility for new investment into the country was regarded as a brazen disregard for even minimal ethical standards.

Fearing that leaving Bhutto in office would bankrupt the state, President Leghari took action on 5 November and replaced her with yet another Pakistani caretaker regime, this time headed by former Senate leader Meraj Khalid. With a promise to hold all corrupt politicians and bureaucrats accountable, elections were scheduled for 3 February 1997. Although the interim government took little effective action against charges of corruption, Pakistan's electorate ultimately levied a high price at the polls for Bhutto's behaviour in office. In a stunning reversal, her party won fewer than a third of the seats. Bhutto's rival, Nawaz Sharif, now has a huge majority in the lower house, controlling 181 of the 217 available seats. After this political humiliation, Bhutto may now choose to distance herself as gracefully as she can from Pakistani politics and await another opportunity for a political rebirth.

Not only did Bhutto's political career suffer, but democracy in Pakistan was also fundamentally challenged by the events of 1996. Bhutto's was the third government in succession to be removed by presidential decree and with the tacit support of the military. After the dismissal of the elected government in November, Khalid's interim government took the extraordinary step of establishing a new CDNS. The Council formally includes the military, and was assigned responsibility for setting national-security priorities, coordinating defence with national and domestic policy, and defining economic and financial policies affecting national security.

The establishment of this non-accountable body, to be chaired by the President, not the Prime Minister, was effectively an admission by Pakistan's élite that democracy had failed to produce leaders who could be trusted with the nation's well-being. A new era of government by élite committee, with the armed services playing a leading role, was thereby ushered in. The high hopes that had accompanied Bhutto's return to power in 1993 had come to a disappointing end.

After Bhutto's tenure, Pakistan's economy is barely hanging on. During the year, inflation continued to rise, while public borrowing – largely to cover corrupt practices – kept interest rates high. Despite an 8.5% devaluation of the rupee, a disappointing cotton crop brought in less revenue than hoped for, while debt servicing and defence expenditures continued to consume at least 60–70% of Pakistan's annual budget. The IMF had promised $600m in support for Pakistan, but the fiscal mismanagement was such that the third $200m tranche was withheld. It was discovered that just before IMF inspectors arrived to evaluate Pakistan's fiscal status, money had been transferred from private banks into the State Bank of Pakistan to make it appear that Pakistan had met the conditions for aid.

Bhutto tried to meet the IMF guidelines for renewed funding by imposing a mid-year tax, but chose a value-added tax which effectively increased costs to the consumer and to business, while ignoring fundamental tax reform in the agricultural sector. The tax provoked a predictably angry response from the business community, as well as short-lived street demonstrations. The continuing refusal of the landed gentry to tax profits on agriculture, in a country which has never undergone land reform and therefore continues to be dominated by a small cadre of powerful feudal landlords, deferred problems rather than solved them.

Once Bhutto was removed from office, a Pakistani Vice-President of the IMF, Shahid Javed Burki, was named interim Finance Minister. He attempted to sort out the nation's finances, and was able to gain an additional IMF commitment of $231m for 1997. Illustrative of the problem was the discovery that the National Bank of Pakistan in Karachi alone had spent 382m rupees on renovation over the previous three years, while an additional R532m had been spent on apparently fraudulent purposes in 1996. The corruption was so pervasive that in September 1996, Bhutto admitted that only 200,000 Pakistani citizens, many of them government employees whose wages were attached, had so far paid taxes that year.

An agricultural tax should be Nawaz Sharif's first priority, but with such a tax affecting the few supporters Bhutto still has, she can be expected to oppose it in the Constituent Assembly. Sharif himself had been removed amid accusations of corruption in 1993 and few Pakistanis can be sanguine that the new administration will actively root out corruption and put Pakistan back on sound financial footing.

National Security

In response to rumours that India planned to hold a nuclear test at the beginning of the year, Pakistan reportedly prepared to conduct its own test. This issue had no sooner died down, however, than the US media reported that China had transferred sensitive uranium-enrichment technology to Pakistan, had provided it with M-11 missiles, and was assisting Pakistan in constructing a missile factory outside Islamabad. Both Islamabad and Beijing predictably denied all such stories, but New Delhi responded with alarm, announcing that there would be no delay in its own missile efforts.

The *Taleban* victory in September 1996 over the various warring factions in and around Kabul in Afghanistan seemed on the surface a success for Pakistani policies as well. Although the Pakistani authorities denied involvement, few believed this, and its Embassy in Kabul was bombed in May 1996 as relations with the besieged Afghan government deteriorated (see maps, pp. 252–54). Without significant support from Pakistan, it is doubtful that *Taleban* could have armed itself as fully as it did or have enjoyed such success in the field. It was not simply that it was well armed; it was disciplined and fighting for a cause against poorly motivated troops. The other factions' ethnically based and sometimes mercenary armies largely melted away in the face of the fundamentalist-inspired *Taleban*.

Pakistan would like to see Afghanistan as a doorway to the new Central Asian former Soviet republics, but it may be difficult to resist the backlash from *Taleban*'s success. Pakistani leaders must find some way of explaining away the power of fundamentalist Islam as a force for national integration. Pakistan has historically been unable to integrate its people around the moderate brand of Islam practised throughout much of the country, and now must be prepared to argue that a more extreme version may be suitable for people supported by Pakistan in a neighbouring state, but is not good for Pakistan itself.

Relations with the United States improved significantly during the year as a result of Prime Minister Bhutto's mid-year visit to Washington. Her arguments persuaded US President Bill Clinton that Pakistan should either be returned the money it had spent on embargoed military goods – particularly the 28 F-16 fighter jets which it had paid for, but never received – or should receive the goods themselves. The administration backed the passage of the Brown Amendment to the September 1985 Foreign Assistance Act which granted Pakistan a one-time waiver of the embargo on military goods imposed in 1990 when then President George

Bush could not certify that Pakistan did not possess a nuclear explosive device. Islamabad was hard pressed to follow up on this diplomatic success, however. The military equipment had begun arriving in Karachi before Bhutto was removed from office, but the third and final shipment was not completed until the end of January 1997. The US had refunded Pakistan some of the money it claimed, but the F-16s remained off limits, their planned sale to a third party was never completed, and the goods that were actually delivered had deteriorated somewhat during the six-year embargo. If nothing else, the transfer of goods at least symbolically restored harmony to US–Pakistan relations.

Improvement in Kashmir

The most significant step forward for India in 1996 may have been the organisation and holding of elections in Kashmir. Kashmir had voted in the April–May general election, but no state-level elections had been held since 1987, when then Prime Minister Rajiv Gandhi tried to manipulate the process. Maintaining democracy has been an important element in allowing Kashmiris a sense of control over their own fate. But without a legitimate forum over the past decade through which to express their preferences and control their state internally, rule from New Delhi was intolerable. According to Delhi, over 1,100 militants were killed in 1996 alone, capping a near-decade of tumult following Gandhi's flawed policy.

In an effort to escape from these policies, elections were conducted in September, despite Pakistan's vehement protests. Claiming that the turnout was 50–60% and that the election was free and fair, India proclaimed victory for the National Conference, headed by Farooq Abdullah, which won 58 of the 87 seats contested. Like his father, who had led the National Conference at the time of partition in 1947 and who had no desire to see Kashmir linked with Pakistan, the new Chief Minister promptly stated that Kashmir was 'part and parcel' of India. He added that he intended to seek greater autonomy for Kashmir, but, at least from New Delhi's point of view, these were subjects for discussion rather than for further violence.

The United Nations recognised the improvement that had taken place in Kashmiri politics by removing it from its list of trouble spots. This understandably enraged Pakistanis, who cited the case of the *Hazratbal* mosque, which had been seized and burned in the spring, as the most egregious example of Indian tyranny. Furthermore, they had agreed in late 1993 at US urging not to push human-rights issues in Kashmir in the expectation that Washington would pressure India to be more forthcoming during talks in January 1994. In the event, the two countries exchanged a few shop-worn arguments, India belatedly presented Pakistan with a set of bland papers on innocuous disputes, and the two sides stopped talking.

Now, Pakistan again feels betrayed on this issue, and the dispute has taken on some of the flavour of the Israeli–Palestinian dispute over Jerusalem's final status. Pakistan insists that Kashmir be placed at the top of the agenda in any future discussions with India, while India insists that, if it is included at all, it be left to the end after less contentious issues are resolved. The two countries will soon have an opportunity to sort out how they will approach the Kashmir question, among other outstanding problems. Nawaz Sharif has proposed a summit meeting, accepted almost immediately by his counterpart Deve Gowda, and senior officials were scheduled to meet before the end of March 1997 to prepare for the talks.

Even with his massive majority in the lower house, Prime Minister Sharif will have difficulties over Kashmir. Pakistan is increasingly a captive of the rhetoric that surrounds the area. For domestic reasons, the issue cannot be ignored by Pakistani politicians, but with at least the beginnings of democratic normalcy returning to the valley, Pakistan's ability to influence events is diminishing. This could have serious security ramifications, however. To the extent that Islamabad sees Kashmir slipping from its influence, it may be less risk-sensitive in its support for radicals and terrorists who want to defy the elected leadership. The Kashmir radicals will want to prevent Chief Minister Farooq Abdullah from succeeding, and may feel driven to desperate action. Indeed, bombs exploded during a speech he made in September 1996, although it was not clear that he was the target. Nevertheless, the security threat against him was made evident when a second bomb attack targeted Kashmiri government ministers in January 1997. If Pakistan is seen to be supporting radical action against Kashmiri leaders, another crisis may erupt and again threaten relations between India and Pakistan.

Preoccupied with Domestic Issues

Although election politics and government-building occupied the Indian leadership during the year, on balance steady economic progress was the primary constant. In the debate that followed the reports of a planned nuclear test, those who argued that a test would undermine economic developments appear to have won the day. Although the election has brought to power a coalition of forces with relatively little in common, it has so far proved able to agree on the key issues of continuing economic expansion and opening the diplomatic door to Pakistan. Ironically, the United Front is now poised to reap the rewards of the economic reform institutionalised by the Congress Party. For its part, as a minority party, Congress has to contend with reduced power at state level, limited access to the perquisites of national power, no new generation of leadership capable of commanding a national mandate, and political marginalisation.

Although leadership in Pakistan has by now become 'rule by committee' in the form of the CDNS, the declining economy and domestic politics in 1996 prevented the leadership from breaking new ground in foreign or defence matters. This may change as the Pakistan Muslim League under Nawaz Sharif assumes power following its strong showing in the February 1997 elections. But cynicism was the main victor in the elections – the turnout was below 30% of eligible voters – and Pakistan could still face further political warfare.

Sharif's huge parliamentary majority may allow him to avoid political back-biting and bring about real reform. But with only a small proportion of the electorate supporting him, Sharif may find it difficult to go beyond draping himself in the symbols of power in Pakistan: Islam, Kashmir and the nuclear programme. To the extent that the politicians become captives of these policy issues, rather than formulating change, the danger of conflict with India increases. Those desiring independence for Kashmir are astute enough to exploit their symbolic leverage, forcing Islamabad to continue supporting a cause that is not entirely to Pakistan's benefit. The success of the *Taleban* is a mixed blessing for Pakistan; although the new Kabul government is partially Pakistan's creation, its insistence on a rigid fundamentalist approach to government, if allowed to develop within Pakistan itself, would further exacerbate the country's regional, ethnic and religious divisions. And so long as the dialogue with India is moribund, the nuclear response will continue to be seen as the only available means to deny India an opportunity to exploit Pakistan's fractious internal politics.

Africa

To non-African eyes, Africa appears a continent beset by catastrophe. Yet in some areas this was far from the case in 1996. In southern Africa, the stabilising influence of South Africa has been critical. Zimbabwe's economy, for example, continues to grow, despite the dissatisfaction of government workers with their salaries. Even the leftist government of Prime Minister Pascoal Mocumbi in Mozambique has been considering increasing the number of key industries it is privatising, and prospects to end the long-running war in Angola seem more positive than in the past.

Democratisation also featured strongly in Africa in 1996 and some key elections were held. Uganda's Yoweri Museveni was voted back as President in April, and Jerry Rawlings was re-elected President of Ghana in October, although his National Democratic Congress Party did less well, illustrating Rawlings' popularity. Zambia's elections on 18 October 1996 were more problematic, with Frederick Chiluba excluding former President and political rival Kenneth Kaunda from running in the presidential race for dubious reasons of family background.

Despite these more positive developments, two overarching trends emerged in 1996 which will significantly influence Africa's future stability. The first is the widening crisis in Central Africa and its strategic implications for the whole continent. The second is the gradual, yet significant, disengagement of the international community from Africa. To most of the outside world, domestic and other concerns are more pressing than constructive engagement in Africa.

The Arc of Conflict and Crisis

A swath of violence stretching from the Indian to the Atlantic Ocean spread across the continent in 1996. Some were long-standing disputes, while other new ones sprang to prominence. The line of conflict runs from the continuing tribal war in Somalia through conflict in Sudan and the ethnic hatred of Rwanda and Burundi to instability in the Central African Republic (CAR). In 1996, Zaire joined this group. The giant of Africa, bordering nine other states, is faltering, battered by years of misrule, the prospective death of its long-standing kleptocratic President, Mobutu Sese Seko, and the serious threat of a rapidly growing rebel war in the east.

While the epicentre of this arc of violence is the Great Lakes region, destabilisation has moved outwards. National frontiers appear no obstacle to conflict, although some pretence is still maintained. Rwanda, for

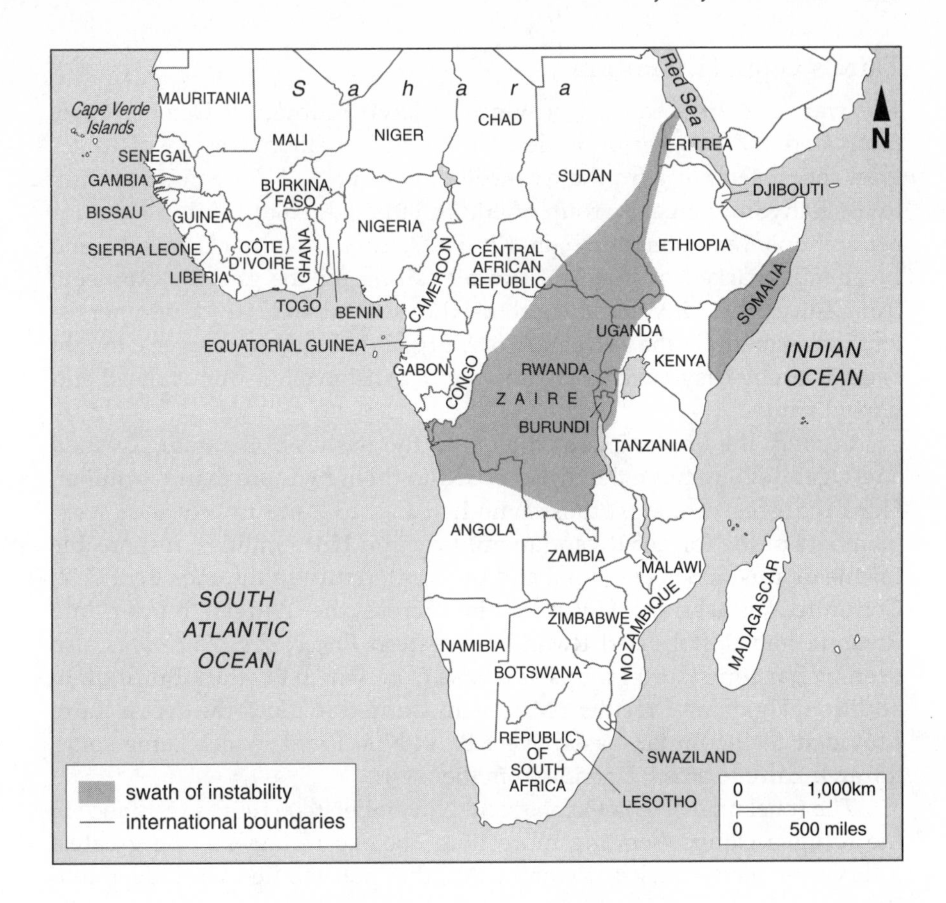

example, claims it is not involved with the Zairean rebels, but inhibitions about crossing borders seem to have relaxed. Instead, rebel groups operating from or supported by friendly countries help destabilise neighbouring states in a kind of 'war by proxy'. These activities have forced neighbouring states into alliances of mutual interest. Sudan, for example, supports Zaire against Uganda: Zaire accuses Uganda of direct involvement in the war in the east, while Sudan maintains that Uganda is assisting the Sudanese groups fighting the Khartoum government on its southern border. Sudan also faces attacks on its eastern borders from opposition groups operating from both Ethiopia and Eritrea. Uganda, on the other hand, is suffering incursions on both its western and northern borders by rebels operating from Zaire and Sudan.

The rebel uprising in the east of Zaire, originally believed to be a local insurrection, now threatens the entire country. But it is also intertwined with the future of Rwanda and Burundi which both face their own internal crises (see map, p. 256).

Zaire's Uncertain Future

The war in Zaire's eastern province of Kivu erupted in October 1996, motivated by two major issues. First, the rebellion was inspired by grievances stemming from the cancellation in 1981 of Zairean citizenship to the Banyamulenge, a group of ethnic Tutsi who had lived in Kivu for generations. In September 1996, Kivu's Deputy Governor, Lwasi Ngabo Lwabanji, sparked off the problem by calling for the group's expulsion from Zaire. The Banyamulenge refused to submit and, faced with persistent harassment by the Zairean Army, urged on by Hutu refugees, fought back, winning easy victories against Zaire's dishevelled, demoralised and unpaid army.

Second, the uprising coincided with the wishes of Burundi, Rwanda and Uganda to remove the threat posed to them by more than 1.5 million Hutu refugees. Amongst those who had fled to Zaire in July 1994 were members of the former Rwandan military and Hutu militias, responsible for the deaths of 800,000 Tutsi and moderate Hutu in the 1994 genocide. Continued attacks by Hutu militias across the Zairean border into Rwanda had destabilised Rwanda's western *Préfectures*. Zaire was also used by hardline Hutu groups as a base from which to attack Burundi. In addition, Ugandan President Museveni wanted to block the access Zaire provided to the Sudan-based West Bank Nile Front, which came south through Zaire to attack Uganda from the west.

The rebel offensive in October and November 1996 routed resistance in the refugee camps, sending more than 600,000 refugees – the greatest influx ever seen – back to Rwanda. Another 300,000 fled west, deep into Zaire. Since 1994, these displaced people, as well as those in Tanzania, had been fed, clothed and housed by the UN High Commission for Refugees (UNHCR) at a cost of more than $1m a day. Little serious attempt had been made to return them to Rwanda, mainly because the hardline Hutu militias and the former Rwandan Army threatened anyone wanting to return home. In July 1996, against the will of the militias, the Burundian military forced the 70,000 refugees in Burundi back to their homes in southern Rwanda. The overwhelming majority were glad to be back.

With hindsight, Zaire's seemingly spontaneous rebel offensive, spearheaded by the Banyamulenge, fitted neatly into a larger framework. These rebels already had strong links with Burundi, Rwanda and Uganda. The core of the Rwandan Army had once been members of the Ugandan Army, and Burundi's Tutsi-dominated military had always retained close links with Rwanda. The Banyamulenge are believed to have been trained and equipped by Rwanda and Uganda. The rebel force, headed by Laurent Kabila, is actually an alliance of four disaffected groups from the provinces of Kasai and Shaba subsumed into the Alliance of Democratic Forces for the Liberation of Congo-Zaire (ADFL). Kabila, originally from Shaba and involved with the anti-Mobutu rebellion since the 1960s, wants the

Figure 1 *Laurent Kabila*

Laurent Kabila was involved in failed uprisings in the Congo in the mid-1960s. At the time a Marxist, he formed the People's Revolutionary Party in 1967 and received some support from Cuba and the Soviet Union. He was unsuccessful at fomenting any serious opposition to Mobutu, however, and spent most of his time in Tanzania and the south Kivu region. There he came into contact with the Banyamulenge and recognised that linking up with them would provide him with support from Burundi, Rwanda and Uganda. They, on the other hand, benefited from his long experience as an opposition leader and his 'pure' Zairean background – Kabila is from the Luba tribe of Shaba province – which would help remove any image of the rebel uprising as Tutsi dominated. Little is known about Kabila's present beliefs beyond his determination to oust Mobutu and his expressed desire to maintain a unified Zaire.

complete overthrow of Mobutu's government. With the ADFL's rapid advance now threatening areas of the country with key resources and power structures, uncertainty surrounding Zaire's future and even its possible collapse has heightened.

Zaire's doom has been forecast for years, yet the country has confounded predictions and staggered on, even while its basic infrastructure and state functions have decayed. Propping up the country is a combination of private enterprise – two cellular telephone companies have largely replaced the state communications system and the collapse of the road system has allowed a handful of private airlines to establish themselves. The church has played a central role in education and welfare and has even doubled as post office and local administrator. Local non-governmental organisations (NGOs) have also provided some services abandoned by the state. Hospitals and schools still function, but entry is based on ability to pay – from the cost of the medicines to the salaries of the teachers. The result is services for the few, but nothing for the many.

As the rebels advance their influence grows. Yet they are unlikely to be able to hold all of Zaire together without forming other alliances. The remnants of President Mobutu's power base and the state of the opposition parties will also be key factors. Mobutu, who for 31 years has ruled as President over a country of 45m, rich in copper, cobalt and diamonds, was in Europe being treated for prostate cancer throughout most of 1996. Until then, he had dominated the political scene with extraordinary guile and ruthlessness, presiding over the country's destruction while amassing a huge personal fortune. Many believe that Zaire's stagnation, particularly the collapse of the transport system and other infrastructure, has been a deliberate attempt to prevent an organised opposition from mobilising.

The economy is now close to collapse. The growth rate in 1996 was a negative 8% of gross domestic product (GDP) and inflation hovers at around 2,500%. The government responded to this crisis by issuing new banknotes. Its latest issue, in January 1997, was in 100,000, 500,000 and 1,000,000 denominations which were immediately dubbed *billets prostate* in an irreverent reference to Mobutu's illness. It is easy to understand why soldiers are so dissatisfied; they receive 120,000 new zaires (NZ) a month while the exchange rate is around NZ150,000 to the US dollar.

Should Mobutu die suddenly, there will be a political vacuum that even by the most optimistic estimates the rebel force cannot hope to fill in the short term. There is no obvious successor to Mobutu and the opposition is hopelessly divided. The political class in general, ranging from the radical opposition led by Etienne Tchisekedi wa Malumba to the Mobutu-supported *Mouvance Presidentielle,* of which Prime Minister Léon Kengo wa Dondo is part, is unable to work together.

The Army has always been considered the obvious force to fill this vacuum, although its poor showing against the rebel advance has cast some doubt. Yet the Army is something of a paradox: it is perceived as Mobutu's tool of oppression, but it is also one of the few national institutions left. With an estimated strength of 50,000, it has been allowed to decline. Unpaid, ill-equipped soldiers have resorted to looting and extortion to support themselves and their families, a feature which characterised their retreat from positions in the east. The only effective force remaining is the approximately 10,000-strong *Division Spéciale Presidentielle,* loyal to Mobutu.

The prospect of the disintegration of the Zairean military with soldiers running amok is worrying but very possible. The appointment in late 1996 of General Mahele Nieko as Chief of Staff was meant to change things. He is widely respected as both capable and incorrupt, but his position has been undermined by a lack of resources without which he cannot reverse the military's decline. Rather than supplying those much-needed resources, Mobutu has looked outside the country for military support. He has approached Morocco, which supplied troops in 1978 to put down the Shaba *gendarme* uprising in 1978, and would clearly prefer to contract individual mercenaries rather than bolster his own Army, which would create a potential threat to his own position.

Yet despite these manifest problems, the overwhelming desire within Zaire is to remain intact. There has been little evidence of secessionist movements. Shaba (formerly Katanga), where two attempts were made in the 1960s and in 1977–78 to secede, has demonstrated its preference for remaining within a federalist Zaire and little secessionist tendency is evident in other provinces. Civil society is vibrant, but remains unable to influence the political class. Elections forecast for mid-1997 look impossible given the spreading war, the lack of a census and the country's devastated transport systems.

Despite Zaire's internal uncertainties, it remains highly sensitive to external influence, particularly that of the US and France. Until the rebellion broke out, the international community had shown little decisiveness over future policies; instead, confusion and vacillation reigned. But Kabila's success has begun to indicate that the ADFL might provide direction and focus that will be supported by outside powers. The most delicate task for the international community will be to facilitate the transition to a new regime involving the Army, Mobutu's supporters, opposition leaders and Kabila's forces. Failure to weave these disparate groups together might hasten a break-up similar to that experienced in Somalia.

Rwanda – a Sustainable Option?

The rebel advances have clearly aided Rwanda's aspirations to maintain a friendly buffer zone in eastern Zaire. But while Rwanda has at times supported Kabila's forces with its own troops, it is likely to stop short of supporting the overthrow of Mobutu's regime. Outwardly, the closure of the refugee camps has strengthened Rwanda's position. It has retained widespread support from the West, particularly the United States, despite criticism of its support for Kabila. Given the effects of the 1994 genocide, and the concomitant destruction of Rwanda's infrastructure, its transformation into a functioning state has been a remarkable achievement. Its performance indicates a genuine desire by key elements of the government, notably Vice-President Paul Kagame, to bring about long-term change.

But the Rwandan government, always supremely confident, has seriously underestimated the effect that the return of more than one million refugees will have. Indeed, it may have unwittingly imported a civil war. The reaction of the ruling Rwandan Patriotic Front (RPF) to assassinations and guerrilla activity by Hutu extremists – now within Rwanda – has been to increase the frequency and severity of their cordon and search operations. These can ultimately only antagonise and isolate the majority Hutu population.

Long-term prospects for Rwanda look bleak. The government is still perceived by its people as a minority government dominated by Tutsi. Instead, Rwandans want all sections of society to be represented, but democratic elections would inevitably return to power the majority Hutu ethnic group. The justice system is also hopelessly overloaded, and the more than 90,000 people imprisoned for genocide will wait years to be brought to trial. They have refused to take part in any plea-bargaining scheme which the government hoped might reduce the numbers in jail. But, most seriously, the return of the refugees has highlighted the acute problem of too many people and too few resources.

Burundi

There is considerable concern that Rwanda's future might ultimately come to resemble that of Burundi. Both have populations of around seven million with similar ethnic mixes: about 15% Tutsi and the vast majority Hutu. Any prospect for peace in Burundi is undermined by the pathological fear each ethnic group has for the other. The Hutu majority is terrified of the Tutsi Army, while the Tutsi are convinced that the Hutu's ultimate desire is to eliminate them in the kind of genocidal attacks seen in Rwanda in 1994. Seen through this dual lens of paranoia, 1996 was an understandably chaotic and violent year.

Past Tutsi strategy has been one of elimination – to move quickly to suppress any Hutu uprising or threat to Tutsi political and economic domination. The ethnic massacres of 1972, 1988 and 1993 met with little opposition and guaranteed the Tutsi continued power. But the tide is turning. Guerrilla activity, mostly led by the hardline Hutu *Conseil National pour la Défense de la Démocratie* (CNDD) under former Interior Minister Leonard Nyangoma – who has links to Zaire's President Mobutu – has made large areas of the countryside unsafe. The death toll is high, with the number of killings in 1996 estimated at 300–1,000 per week.

In July, Burundi's 'slow-motion coup' finally ended when former Tutsi President Pierre Buyoya ousted Hutu President Sylvestre Ntibantunganya. This destroyed the last remnants of Hutu standing and any respect for the 1993 election results, when the majority Hutu gained power for the first time since independence and President Melchior Ndadaye was elected. After his fall, Ntibantunganya fled to the US Ambassador's residence where he took refuge throughout 1996.

Although generally considered a moderate, Buyoya is constrained by the will of the Army and Tutsi opposition groups. Loss of support from either – in reality closely linked, as the Army is almost exclusively Tutsi – would result in his downfall and possible replacement with a more hardline alternative. His room for manoeuvre is, therefore, tight, and, although he has talked about it, he has yet to prove his sincerity to return to more representative government.

Immediately following the coup, former Tanzanian President Julius Nyerere, who had been leading peace negotiations, convinced other East African countries to join in imposing sanctions. This had two main effects. First, it increased Tutsi fears and feelings of isolation. Second, although the sanctions have been widely circumvented by Burundi's neighbours, coffee exports, Burundi's main source of foreign exchange, have slowed to a trickle, and food and imported goods have doubled or tripled in price. With the fall in revenue, the government might default on paying civil servants and the armed forces which is likely to result in widespread chaos and the fall of the Buyoya regime.

Sanctions, which have worked against other regimes, may be aggravating the situation in Burundi and sowing the seeds of worse violence. The international community has clearly played its best card first. Now it is unwilling to remove the sanctions unless Buyoya grants the kind of concessions that might topple his regime, yet it is similarly unwilling to offer any new 'carrots' to ease the Tutsi perception that the outside world is against them. The external powers are left with few levers of influence. The only positive note was the appointment of the skilled negotiator Mohammed Sahnoun in January 1997 as both the UN and Organisation of African Unity (OAU) representative for the Great Lakes.

Angola – Struggle for Power and Wealth Minus Ideology

The neighbouring country most affected by Zaire's future is Angola which continues to struggle with its long-standing civil war. *União Nacional para a Independência Total de Angola* (UNITA) leader Jonas Savimbi has close links to President Mobutu and northern Angola has proved an important stronghold for his movement. Zaire has given Savimbi access to diamond fields in Angola – thought to have earned him over $500m in 1996 – as well as a staging post for military operations. It is difficult to envisage Savimbi relinquishing control of this strategic area, and by maintaining his position he can remain a potent and destabilising force. The 15,000 former UNITA soldiers who have made their way into UN demobilisation camps during the year are starting to drift away (see map, p. 257).

The latest peace deal, brokered chiefly by Alioune Blondin Beye, the UN Secretary-General's Special Representative, and supported by the third UN Angola Verification Mission (UNAVEM III), is based largely on the 1994 Lusaka Accords. A unified UNITA and Angolan government, headed by José Eduardo dos Santos, has been established and a unified national army of 90,000 is planned which will include more than 26,000 UNITA troops. Some of UNITA's top generals have joined the Angolan Army and other leaders are scheduled to sit in the new parliament. Although the new deal tries to avoid a 'winner-takes-all' option that would exclude UNITA, the main stumbling block remains Savimbi, who has rejected any formal role in the government and had to be prompted by South African President Nelson Mandela to cooperate. It has become clear that the arms handed in by UNITA troops when they entered the demobilisation camps have been barely serviceable, and Savimbi is suspected of still commanding a well-equipped and financed army of more than 10,000, enough to wage considerable havoc on Angola's fragile peace.

Sudan

Sudan, the other giant of Africa on Zaire's north-eastern border, faces opposition on three sides. Egypt accuses it of terrorist activities, and fighting which had been confined to the south has now erupted on a new eastern front. On 12 January 1997, opposition forces moved into the Blue Nile province from Ethiopia taking key towns and threatening the Er Roseires Dam which supplies most of Khartoum's electricity. Until then, cross-border attacks had been launched from either Eritrea or Uganda, not Ethiopia, and the attacks surprised the ruling National Islamic Front (NIF). The government called on volunteers in Khartoum to join the Army to fight the threat, but, reflecting their dissatisfaction with the current government, few joined up.

Sudan now faces a double threat from an alliance of opposition in both north and south. The opposition is made up of John Garang's Sudan People's Liberation Army (SPLA), traditionally from the south, and Sudan Alliance Forces (SAF), under the command of Abdel Aziz Khalid Osman. The US, which has consistently opposed the current Sudanese regime, has been supporting the opposition and, in October 1996, it was reported to have donated $20m-worth of additional military support to Eritrea, Ethiopia and Uganda (see map, p. 258).

Somalia

Left to its own devices since 1995, when the UN and US withdrew their troops, Somalia shows little sign of progress. On 1 August 1996, leader of the Somalia National Alliance and self-proclaimed President of Somalia, Mohammed Farrah Aideed, died from bullet wounds received a few days earlier. For a short period, it was thought that his death might clear the way for a peace settlement. But in a surprise move his 35-year-old son, Hussein Aideed, was elected to take his place. The young Aideed had served in the US Marine Reserve and arrived in Somalia with the US Unified Task Force (UNITAF) in the latter part of 1992. He quickly pledged to continue his father's policies, and claimed the title of President for himself.

Lined up against him are the forces of Osman Arto, formerly Mohammed Farrah Aideed's right-hand man, who now leads an opposition south of Mogadishu, and President Ali Mahdi Mohamed, who leads a group in north Mogadishu. During a meeting in Ethiopia in January 1997, the two forces joined with other Somali clans to establish a new alliance with a 41-member National Salvation Council. Considerable logistical and practical difficulties remain, including such basic questions as where to site the new administration. This latest formulation is a rehash of a similar proposal in 1993 when the UN was present, this time backed by neighbouring Ethiopia. Despite the agreement, Hussein Aideed's absence was crucial and the intractable war broke out again in December 1996, filling Mogadishu's hospitals with wounded.

Nigeria – Business as Usual

International apathy inevitably eases the pressure on countries such as Nigeria and Kenya to improve their approach to human rights and corruption. Nigeria's indifference to such international demands was highlighted soon after its suspension from the British Commonwealth after the execution of Ken Saro-Wiwa at the end of 1995. While the Commonwealth Ministerial Action group was recommending boycotts against Nigeria in April 1996, Nigeria approached China for investments of over $500m in its railway system.

Nigeria's immediate political prospects remain largely unchanged. The presidential election is scheduled for October 1998, and there are continuing signs of divisions in the country, including bombings and disenchantment in the south-west, the homeland of Moshood Abiola who won the last presidential election in June 1993 but who has been languishing in jail rather than ruling the country. The Army, always the greatest threat to the ruling élite, has thus far remained united behind the regime; its officers are rotated regularly as a safeguard against factionalism.

Nigeria's economy looks bright and its future is even brighter. The policies introduced by Finance Minister Anthony Ani have brought inflation down to below 30%, and the growth rate is around 4%. But he has been stopped from going far enough by General Sani Abacha's military government. Ani had planned to abolish the two-tier preferential exchange-rate system, as long-requested by the International Monetary Fund (IMF) and the World Bank. This action, however, would have seriously disadvantaged the privileged position of the Army by limiting its spending power. Increased privatisation of gas and oil operations has also been stalled. Foreign oil companies nevertheless continue to make good profits, and there is a prospect of tapping new oil reserves – a further 500,000 barrels a day are expected to flow from newly exploited deep-water sites. Together with other expansion, this could make Nigeria in the next five to six years the third or fourth largest oil producer in the world, increasing both its economic strength and its regional muscle.

Kenya, too, seems less inclined to bow to international pressure. In January 1997, President Daniel Arap Moi re-appointed Nicholas Biwott to his cabinet as Minister of State. Five years earlier, Western governments had pressured Moi into sacking Biwott, who had been involved in a series of scandals as Minister of Energy and had been implicated in a murder case. His re-appointment came on top of further international allegations of corruption and human-rights abuses. Moi, however, was more interested in internal than external politics, and Biwott's presence in the cabinet strengthens the government by easing tribal rivalries, particularly with respect to the Kikuyu with whom Biwott has close links. President Moi clearly feels confident following the meeting of Kenya's main donors in Paris in April 1996 that released $730m that it had been withholding. Divisions among the ranks of foreign donors have also benefited Moi. The French and Japanese, for example, have recently signed critical energy deals with Kenya, from which others have abstained.

Sierra Leone – A Tentative Success

International apathy may also turn one of Africa's home-grown success stories into a more familiar tale of disaster. In February 1996, Sierra Leone held its first election in nearly 29 years. It returned a democratic parliament with Ahmed Tejan Kabbah as President. The elections were held despite the continuation of a five-year civil war and the opposition of the Revolutionary United Front (RUF). The electorate, which voted amidst gunfire around the centres of Bo and Kenema and even in the capital Freetown, thus ended the five-year military regime headed since January 1996 by Maata Bio who had taken over after deposing Captain Valentine Strasser. Although the military had shown signs of wanting to postpone the elections, claiming that peace was needed first, it buckled under pressure from the growing and more confident civil society backed by international voices, and bullying from former UN Under-Secretary-General James Jonah who led the Freetown-based Electoral Commission.

The success of elections undermined the RUF, weakened by the heavy military losses it suffered at the hands of Executive Outcomes (EO), a South African-based security firm contracted by the Sierra Leonean government. Hiring EO gave the government a hard edge to boost its faltering and corrupt military forces, which were suspected of colluding with the RUF and had been implicated in atrocities against civilians. EO's military strategy included using attack helicopters, radio and night-finding equipment to support a small group of less than 200 ground forces, former members of South African special forces that had been active fighting in Angola and Namibia. With close links to the local population in the diamond-rich eastern region of Koidu, they made good use of local intelligence.

Within a few months, the diamond areas around Koidu had been cleared, limiting the RUF's access to resources and allowing more than 300,000 refugees to return home. Larger mining companies had also re-established their operations, including Branch Energy, whose close connection with EO led to speculation that EO had been granted some diamond concessions as part of its pay. The combination of military defeats and the loss of political initiative brought the RUF out of the bush and into peace talks held in neighbouring Côte d'Ivoire. After protracted and difficult negotiations, a peace agreement was signed on 30 November 1996 which included, at the insistence of the RUF, the withdrawal of EO. Despite this agreement, there has been continued sporadic fighting, although many RUF fighters have now emerged from their bush hideouts.

Even with this promising start the country has been unable to attract international attention and finance. A mission of UN observers and peacekeepers has been approved by the UN Security Council, but will not arrive until mid-1997 at the earliest. There is real danger that the optimism that accompanied the elections will falter without the comprehensive demobilisation and reintegration of both the Army and the RUF. International support for the economy is also vitally needed.

International Indifference

If there is a common thread running throughout Africa it is fading international attention. The outstanding feature of Western policy in Africa is its absence. If the situation inside Zaire looks uncertain, it is not helped by the lack of any clear international position over the last two years. When more than one million refugees arrived from Rwanda in 1994, Zaire agreed to accept the presence of foreign troops as the precondition to a large-scale humanitarian operation. The operation resuscitated Mobutu's political fortunes. As Zaire faced collapse in March 1997, and with few options available, France in particular has supported Mobutu in a bid to hold the country together. Yet this policy has little coherence, for while France is concerned that Zaire might fragment without Mobutu, it also knows that until he goes there will be little opportunity for change. Others have grown to accept Mobutu as 'history' but have no clear policy for how to deal with a probable successor.

The US position is uncertain, although there is growing suspicion that it has not been merely an impartial observer. The US has close relations with Rwanda and Uganda, and both countries have done much to help train and equip Kabila's growing forces, which has contributed greatly to his continued successes. Without any real assets to bring about change within Zaire, the US appears to believe that Kabila's efforts are the only realistic way to either bring Mobutu to the table, or to depose him. Suspicions concerning US involvement aggravate France's concern at its loss of influence and the possible impact on other Francophone countries in the region. The suspicions are mutual, and at times open antagonism has been a feature of relations between France and the United States over Central Africa.

Yet France also appears unwilling to become more involved and has announced plans to scale back its military forces in Africa. Already its traditionally interventionist role in Africa is becoming less decisive following a series of setbacks in old areas of influence. There have been mutinies in Congo and Guinea and problems in the CAR. France dispatched troops to the CAR in April 1996 when the elected government of Ange-Félix Patasse was threatened by an Army mutiny, triggered by its dissatisfaction at its general treatment and the specific complaint that soldiers had not been paid. The French operation has now become overwhelmed. To help quell the revolt and keep French soldiers out of danger, the French military budget is now funding salaries for the CAR military. But the determination to act quickly, and clearly, to support a political leadership seems to have eroded, and with it France's earlier capacity to dictate domestic affairs in many African countries.

With all the major Western countries wanting to disengage, or losing influence, South Africa has been pushed to take a more active role. It has responded by taking a lead in peace negotiations. In early 1997, it invited the Zairean rebel leader Kabila to Johannesburg to meet Zairean officials. It has also taken more interest in trying to help settle the war in Angola, a war in which the former apartheid regime played a direct interventionary role.

It is difficult not to believe that the Western policy of 'promoting African solutions to African problems' is also a means to disengage further from the continent. The support for sanctions imposed by regional governments against Buyoya's government in Burundi reflects more the support for East African states than overwhelming support for the policy. Former US Secretary of State Warren Christopher, during a visit to Africa in October 1996, reiterated the US offer to contribute $20m out of $40m to fund a standing force of 10,000 African soldiers with a UN mandate to create safe havens for civilians in areas of crisis. But Christopher's trip, just before the US elections, was seen more as helping politics at home than as a genuine attempt to engage constructively.

Help is Needed

What happens in Zaire will have a profound effect on the whole of sub-Saharan Africa. But the long-standing and legitimate grievances of its people have not awakened the conscience of the international community which is less and less inclined to become involved. International resolve may develop, but perhaps too late to influence the problem, or possibly to stave off catastrophe. Instead in Zaire, the advance of the rebels has become *de facto* policy, a hands-off approach where the hope is that war might bring better things. It is risky in the extreme.

The test of whether the West can maintain its indifference will come if Mobutu is deposed and Zaire collapses. This is a country the size of Europe, rich in oil, diamonds and other mineral resources. Should the West allow it to descend into chaos without attempting to shore up a reasonable, unifying government there will be little hope for any other African country facing an uncertain future.

South Africa – Waking Up to Reality

In 1994, South Africa escaped what many observers believed was its manifest destiny: a racial conflict of apocalyptic dimensions. Much of the following 18 months was devoted to a celebration of that escape and the political miracle that had brought it about. The whole country, its political leaders, sportsmen and businessmen basked in South Africa's new reputation as an enviable example of sanity and reconciliation. Most of the celebrations focused on President Nelson Mandela – the charismatic symbol of the country's reconciliation – reaching a peak during his state visit to London in June 1996 when 100,000 filled Trafalgar Square to hear him speak.

But even as a South African President received this extraordinary homage, the harsh realities of normal life began to signal that the time for festivities had passed. There was general agreement that the new sense of nationhood was not yet so entrenched that a deeply divided society could survive without the Mandela 'glue' that bonded it together. Nevertheless, 1999 – the year set by Mandela himself for his retirement – was no longer seen either in the country or outside as a feared unknown, but as the inevitable next step in South Africa's transition to a normal society.

Problems Galore

Despite almost universal respect for Mandela's personality and achievements, there was impatience in 1996 with his sometimes Quixotic defence both of incompetent lieutenants and of old allies such as Libya's Moamar Gadaffi, his increasing intolerance of press criticism and his occasional maladroit intervention in foreign and domestic policy. All this contributed to, rather than resolved, some of South Africa's more pressing problems, including a loss of investor confidence which by the end of the year had wiped 25% off the dollar value of the rand.

Foreign investors, who had been responsible for a R21 billion capital inflow – mainly into equities – since 1994, were responding to a cocktail of issues. These included an explosion of violent crime, which propelled South Africa to the top of the international league table with a murder rate five times higher than the US and six times higher than Russia. Other factors were: the abrupt departure of Minister of Finance Chris Liebenberg in April 1996 and the appointment of an untried African National Congress (ANC) politician, Trevor Manuel, in his place; bureaucratic incompetence which brought most of the government's promises to the poor to a standstill; and some high-profile examples of political corruption. Investors were further disturbed by the ANC's manifest unwillingness to deal with the country's militant trade-union movement over the privatisation of state assets and an inflexible labour market.

In the Grip of Crime

If foreign investors had been slow to respond to South Africa's opening to the outside world, international criminals from Chinese Triads to Colombian drug cartels and Nigerian money-launderers were not. They quickly seized the opportunities offered by a hungry country awash with guns, policed by an undermanned, under-trained, demoralised and often corrupt police force, overflowing prisons and a criminal justice system that had swung from being excessively punitive to overly permissive (bail was seen as a right for those awaiting trial). Anger over South Africa's lawlessness was not restricted to the country's wealthier suburbs, although the casual killing both of the Chief Executive Officer of a major multinational company and the father of a leading banker, and the mugging of the President of the Constitutional Court Arthur Chaskalson in September 1996, probably did more harm to the country's economic prospects than the whole experience of crime in Soweto.

The problem was exacerbated by the government's initial refusal to respond to almost universal concern over violent crime. In an opinion poll, 70% favoured the return of the death penalty. Gradually, however, the issue of crime crept up the government's agenda, helped by a private-sector organisation, Business against Crime, which brought in management consultants to unearth and address some of the most urgent policing problems. These included the absence of any central police data bank, the fact that many policemen were illiterate, others could not drive and many were themselves guilty of violent crimes.

In early 1997, Police Commissioner George Fivaz conceded that the crime rate in the police force was higher than in society as a whole. In his March 1997 budget, new Finance Minister Manuel announced a 15% increase in funding for the police, the justice system and the prison service. Correctional Services Commissioner Khulekani Sithole suggested that disused mine shafts be turned into maximum-security prisons for violent offenders, and the Army was deployed to assist police in urban areas.

The Effects of Reverse Discrimination

The problems in the police service – although compounded by the legacy of the apartheid years when the police were little more than the main instrument of state repression – were repeated throughout the public sector. A policy of fast-track black advancement to senior levels put strain on the treasury which had to provide massive redundancy packages for old-guard, white civil servants. An unintended consequence was that the best and the brightest white bureaucrats took their golden handshakes and then hired themselves back to government departments as 'consultants' to their successors, doubling the departmental pay bill. The worst of these bureaucrats, facing an uncertain future outside public service, clung to their jobs, thus increasing the slide to inefficiency and administrative delay. In a landmark decision in March 1997, however, the government's affirmative-action policies suffered a setback when the Pretoria High Court declared the appointment of 30 state attorneys unconstitutional as only women and blacks had been considered for the posts.

The policy aggravated an already acute skills shortage caused by decades of educational neglect in the black townships. This, in turn, was exacerbated by an education policy which led to more than 10,000 teachers accepting redundancy rather than redeployment from predominantly white schools to understaffed schools in the townships and rural areas. In early 1997, the government itself accepted that the policy had been misguided – but the damage had been done.

An official bias against white males in the public sector and the threat posed by violent crime led increasing numbers of young professionals to move abroad, resulting in net official emigration for the third year in a row. At the same time, a more potent source of instability was the exponential growth in illegal immigration as Africa's poor, desperate and dispossessed from as far afield as Nigeria, Rwanda and Sierra Leone continued to flood across South Africa's porous borders. By definition, it is impossible to provide accurate figures for this influx, but official estimates range from 2–8m. In a country with a 33% unemployment rate which has not yet begun to meet the needs of its own population – increasing naturally by one million a year – illegal immigration could seriously strain relations with South Africa's neighbours and damage its own delicate social fabric. The fact that many of these migrants are easily identified by language and most compete as hawkers and street traders with South Africans on the lowest rung of the economic ladder has already led to violent clashes.

Other Economic Problems

In the formal sector, too, South Africa continued to face rising joblessness. Despite a GDP increase of 3%, formal-sector employment in the first three months of 1996 declined by 52,000 while wages increased by 12.3%. South Africa is one of the most inflexible labour markets in the world, languishing at a mere 11 out of 100 on the international competitiveness scale, reducing its attractiveness to foreign investors in a world where international competition for investment has never been higher.

The paradoxical combination of high wages, low skills and massive unemployment is both a legacy of the apartheid era and the product of a militant trade-union movement which, having contributed greatly to the struggle for political liberation, forms an important if awkward part of the ANC's constituency. Although only 25% of South Africa's labour force is unionised, the flexing of union muscle, particularly large-scale labour disruption in the public sector, was initially enough to raise a question mark over the government's willingness to privatise South Africa's giant utilities and other state industries.

The pace did not quicken even after President Mandela insisted that the privatisation policy was non-negotiable. Nevertheless, by the beginning of 1997 moves to sell off 30% of Telkom, the country's telecommunications utility, were proceeding, albeit slowly, and there were hopes that union opposition could be eroded by involving union members as shareholders. There was somewhat less hope that the unions could be persuaded to exclude small firms from wage agreements, thus allowing the labour market to expand.

Delivery of key elements in the ANC's much-vaunted Reconstruction and Development Programme (RDP) – particularly housing and basic infrastructure in the townships and squatter camps – would have been a much-needed spur to labour-intensive economic growth. Housing construction, however, was frustrated by bureaucratic inefficiency and, more critically, by the culture of non-payment of rent, interest and service levies which had flourished as part of the campaign to make South Africa ungovernable during the apartheid era. This has since become part of an apparently ineradicable culture of entitlement among the majority population which remains deaf to persistent pleas from Mandela and other leading politicians to end the rent boycott. The result was that by mid-1996, only 11,000 new houses had been built for the poor since 1994. This was a lower rate than under the previous government and, in an effort to breathe new life into the programme, the RDP office was dismantled and its duties taken over by the office of Deputy President Thabo Mbeki.

The culture of non-payment has, however, done more than kill off large parts of the RDP and hamper economic growth. It has effectively bankrupted many local authorities (the giant electricity utility Escom claims that it is owed more than R1.3bn in unpaid bills) and has led to riots on university campuses as hard-pressed authorities attempt to recover fees from students. In early 1997, riots in a predominantly 'coloured' township on the East Rand injured 200. Residents claimed that authorities did not extend the same forbearance to them as they had to black townships over the non-payment of rents and levies.

The Impact on Politics

Anger over the crime rate and tales of government corruption, as well as a growing crisis of defeated expectations, could undermine the ANC's authority in the long term and fuel the political hopes of dissidents such as Bantu Holomisa. Expelled from the ANC on 30 August 1996 for whistle-blowing on alleged internal corruption, he began to attract large crowds to his rallies in the Eastern Cape. Nevertheless, on the macro-political level, South Africa seems more likely to follow the example of other African states where opposition parties wither away or are absorbed into a coalition government and where opposition is expressed within the main political movement, not outside it. This is despite the fact that the Government of National Unity (GNU) virtually disintegrated in mid-1996 when Deputy President F. W. de Klerk took the National Party (NP) out of government and into opposition. The other partner in the GNU, the *Inkatha* Freedom Party (IFP), led by Chief Mangosuthu Buthelezi, remained in the government, but to little effect.

Subsequent attempts by de Klerk to subsume the National Party into a broad-based new opposition, including both the IFP and the liberal Democratic Party, failed – partly because of strenuous resistance from the National Party in the Western Cape, it's only remaining power base. There are doubts as to whether the NP, either in its old or new incarnation, will long survive at the national level. Meanwhile, the Democratic Party has resisted both de Klerk and an invitation from President Mandela for it and a re-constituted Pan Africanist Congress (PAC) to join the government. Although the IFP remains the majority party in KwaZulu–Natal, its grip is weakening and some commentators doubt that it would survive Buthelezi's death or retirement for long.

A clear indication of the marginalisation of Buthelezi and *Inkatha* can be found in the provisions of the new Constitution, negotiated over a two-year period, and finally signed into law on 4 December 1996. The text had been approved almost unanimously by the Constitutional Assembly, but it was passed without the participation of the IFP which continues to demand greater regional autonomy. The Constitution is far from a federalist document, as Buthelezi wanted, but it does give the nine provinces somewhat more powers. The previous upper house, the Senate, will be replaced by a Council of Provinces, which will give the provincial premiers a more direct say at the national level. But the provinces will receive no new powers either to collect their own taxes or to control the police forces. On other matters, the Constitution, with its heavy emphasis on equality and an embedded, and strong, Bill of Rights, is a document which moves South Africa under the ANC from a country that guaranteed a place in government for the minority white party to a government based on majority rule.

Although the ANC overshadows all other parties on the national political stage, it is not without divisions in its own ranks. Political infighting and allegations of maladministration have led to the removal of three provincial leaders, including the effective and admired Patrick Lekota as Premier of the Free State where resentment at the centre's interference in local politics continues to bubble. At the same time, Winnie Mandela, the President's divorced wife and a one-time ally of Bantu Holomisa, has regained the leadership of the powerful ANC Women's Movement where she continues to pose a threat to party moderates.

It was, however, the moderates, particularly in the Treasury and Industry team who scored the major successes of the year. Led by new Finance Minister Manuel and Industry Minister Alec Irwin, they embodied a sense of fiscal discipline and prudence. Despite the fact that the rand had become substantially undervalued, by the end of 1996 they had restored some of the confidence of overseas investors and improved economic prospects.

Economic Plans and Some Recovery

The new sense of economic leadership was signalled in a September 1996 government policy document, 'Growth, Employment and Redistribution' (GEAR) which provided guidelines for fiscal management and industrial policy. The central tenet was a commitment to cut the budget deficit from 5.4% to 3% by 2000. Its thesis – eagerly embraced by the private sector but with less enthusiasm by the trade unions – was that job creation was the best method of re-distribution and should be left to the private sector operating in an enabling environment.

GEAR's immediate target was to create 270,000 new jobs a year over the next five years, three-quarters of which would be generated in the formal and informal private sector in an economy growing at an average rate of 4.2%. It also anticipated an annual inflow of US$509m a year in direct foreign investment. Its longer-term aim is an annual GDP growth of more than 5% from 2001, and an annual employment growth of 500,000 jobs a year. Key to its success is labour-market flexibility, especially in the small-business sector, and the government's commitment to lowering South Africa's high tariff barriers as its embraces the rules of the World Trade Organisation (WTO) – neither of which are popular with the labour unions.

Although, given the strength of union opposition, GEAR is not much more than a wish list, Manuel continues to demonstrate the fiscal discipline necessary if the government is to cut the budget deficit. In his March 1997 budget, he announced that the initial target of a 5.1% budget deficit had been met and would be cut to 4% in financial year 1997. Defence had its budget slashed by 9.4% in real terms, while there were substantial increases for housing and the police force, poverty-relief programmes and aid for pensioners. At the same time, Manuel announced a further relaxation of foreign-exchange control for residents – a necessary move to boost investor confidence and essential if South Africa is to qualify for an IMF standby facility.

Meanwhile, the competitive benefits of a cheaper rand have also begun to flow back into the economy, stimulating manufacturing industry and exports which in January 1997 chalked up a 50% increase compared to January 1996 and helped South Africa realise an annual trade surplus of $2.5bn. Nevertheless, the growth rate remains at 3% and inflation at 9%, and although the education budget represents 45% of all social spending, concentrating on primary and secondary education has meant huge cutbacks in university funding. This has provoked widespread discontent and persistent fears that South Africa will never be able to bridge the skills gap and boost its performance against its main competitors – the 'tiger' economies of East Asia. South Africa increasingly looks to two such nations for investment. Malaysia, particularly, is emerging as a major partner in a number of key South African enterprises. The Malaysian example of a corporate state with close links of patronage and purchase between government and the private sector could also prove a seductive model for South African politicians.

Foreign and Defence Issues

If South Africa assiduously cultivated its economic ties with the Far East and the Pacific Rim in 1996, its relations with the West, especially the United States, and its immediate African neighbours were less sure-footed. President Mandela's desire to use his international standing as a moral arbiter faltered in a policy muddle over the crisis in Nigeria. A later, much-publicised effort supported by the United States to bring Zairean rebel leader Laurent Kabila and President Mobutu Sese Seko's envoys together in South Africa in February 1997 failed. As President of the 12-nation Southern African Development Community (SADEC), Mandela has also had talks with the warring factions in other African trouble spots such as Sudan and Swaziland, and was ready to send South African troops to help the Lesotho government quash a police mutiny.

Mandela is determined that Africa should dispense with Western intervention to resolve its periodic crises and establish its own multinational peacekeeping force. To this end, a combined peacekeeping exercise involving troops from Botswana, South Africa, Tanzania and Zimbabwe was announced in March 1997.

Military cooperation represents a rare success for SADEC, whose members resent South African dominance, particularly trade imbalances, currently ten to one in favour of South Africa which nevertheless maintains tariff barriers against Zimbabwe textiles. Efforts to establish a free-trade area and boost prosperity throughout the region were boosted by South Africa's lengthy – but finally successful – negotiations to achieve a favourable trade deal with the European Union. On 24 March 1997, it was announced that Spain would not block South Africa's bid to join the EU's Lomé Convention as it had threatened to do. South Africa's entry into the Convention is expected to be ratified on 24–25 April when the EU is to meet with its African, Caribbean and Pacific partners.

Equally lengthy negotiations with the United States to relax its ban on South African arms sales – kept in place because of arms smuggling by Armscor during the sanctions era – were more successful and the six-year dispute was finally resolved in January 1996. The ban had effectively frustrated a South African bid to supply its *Rooivalk* helicopter to the UK and had denied Armscor's successor, Denel, access to US technology and international markets.

At the time the deal was announced, however, South Africa tried US patience by revealing that it would bid for a $640m contract to supply tank fire-control systems to Syria. The US Department of State promptly warned that under US law prohibiting arms sales to countries that supported terrorism, US aid of more than $110m to South Africa would be jeopardised. Denel, which boasts exports of more than $300m a year and employs 50,000, was pushing hard for the deal. But, despite statements by Deputy President Mbeki that South Africa would not capitulate and was determined to introduce the principle of 'equivalence' to its relations with the Middle East, the deal seemed unlikely to go ahead. It was reported in February 1997 that South Africa was contemplating a major arms deal with Beijing.

The desire to follow an independent foreign policy was linked to a sense of loyalty to old ANC supporters and a distrust of the West and 'Eurocentric' values. It has also led to closer ties with Iran, including the lease of oil-storage facilities, and statements by President Mandela supporting Cuba and Libya – none of which reassured nervous US investors.

From a Discredited Past ...

The past continued to haunt the present within South Africa, as well. A series of criminal trials and the Truth and Reconciliation Commission (TRC) began to uncover the often horrifying secrets of the dirty war waged by security police and military intelligence against the ANC and other liberation movements during the apartheid era. As more and more former policemen, Army officers and members of the bizarrely named Civil Cooperation Bureau (CCB) applied for amnesty, the country was appalled by a detailed history of atrocities, including the circumstances of the death of activist Steve Biko. The activities of the TRC were not uncontroversial, however, as the ANC appeared reluctant to subject its own less savoury activities to the Commission's scrutiny, while the Commission had to face allegations of racism from within its own ranks and from victim's families, including Biko's widow, protesting at its extension of amnesty to the killers.

One of the most disturbing disclosures followed the arrest for alleged drug dealing of Dr Wouter Basson, who, it was revealed had been involved in the CCB and the so-called Third Force which had come close to wrecking South Africa's peaceful transition to majority rule. He had also been a key figure in the apartheid government's chemical- and biological-warfare programme.

It was subsequently revealed that Dr Basson, who had been dismissed from the Army by former President de Klerk in 1993 following an internal report, had been re-hired as an Army doctor by the new government two years later, ostensibly to secure his silence. Dr Basson had since visited Libya, raising uncomfortable questions about a strange fit between the alliances of the old South Africa, which had actively sought allies among similar international outcasts, and the new. Equally disturbing was the lack of government candour on this and other issues – particularly those relating to allegations of corruption – in a country that now placed the highest value on official transparency. Instead, both Mandela and his deputy and anointed successor Mbeki made no secret of their mounting irritation with a reinvigorated press.

... To a More Hopeful Future

Despite the seemingly intractable problems that have begun to emerge as the glow of South Africa's apparently miraculous transition has worn off, the country has continued to surprise its critics. During the difficulties of 1996, it succeeded in negotiating and agreeing a new Constitution to come into effect in 1999. And even among dire warnings of potential unrest, local elections took place peacefully, ethnic violence is diminishing and the country is now a full working democracy.

The past, with its disclosures of apartheid atrocities, shortage of skills, habits of lawlessness and persistent economic distortions, continues to cast a long shadow over South Africa. Nevertheless, the country is trying to come to terms with this undesirable past while at the same time embracing policies for the future to help it avoid the ideological traps which have successfully beggared much of the rest of post-liberation Africa.

Strategic Geography 1996/97

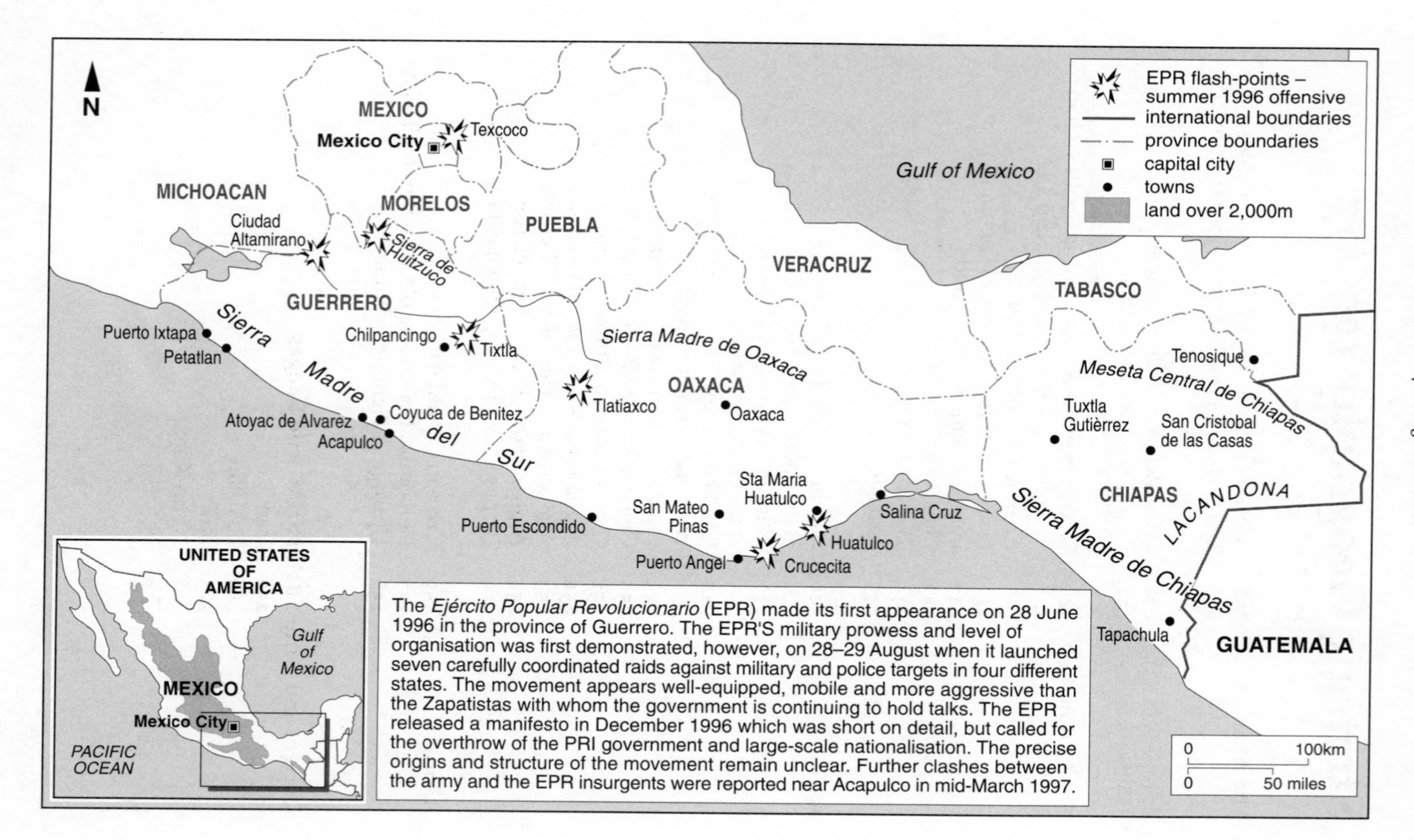
EPR flash-points – summer 1996 offensive
international boundaries
province boundaries
capital city
towns
land over 2,000m
N
MEXICO
Mexico City
Texcoco
MICHOACAN
MORELOS
PUEBLA
Gulf of Mexico
Ciudad Altamirano
Sierra de Huitzuco
VERACRUZ
TABASCO
GUERRERO
Sierra Madre del Sur
Puerto Ixtapa
Petatlan
Chilpancingo
Tixtla
Sierra Madre de Oaxaca
Tenosique
Meseta Central de Chiapas
OAXACA
Tlatiaxco
Oaxaca
Atoyac de Alvarez
Coyuca de Benitez
Acapulco
Tuxtla Gutièrrez
San Cristobal de las Casas
Sta Maria Huatulco
San Mateo Pinas
Puerto Escondido
Salina Cruz
CHIAPAS
LACANDONA
Huatulco
Puerto Angel
Crucecita
Sierra Madre de Chiapas
Tapachula
GUATEMALA
UNITED STATES OF AMERICA
Gulf of Mexico
MEXICO
Mexico City
PACIFIC OCEAN
The *Ejército Popular Revolucionario* (EPR) made its first appearance on 28 June 1996 in the province of Guerrero. The EPR'S military prowess and level of organisation was first demonstrated, however, on 28–29 August when it launched seven carefully coordinated raids against military and police targets in four different states. The movement appears well-equipped, mobile and more aggressive than the Zapatistas with whom the government is continuing to hold talks. The EPR released a manifesto in December 1996 which was short on detail, but called for the overthrow of the PRI government and large-scale nationalisation. The precise origins and structure of the movement remain unclear. Further clashes between the army and the EPR insurgents were reported near Acapulco in mid-March 1997.
0 100km
0 50 miles

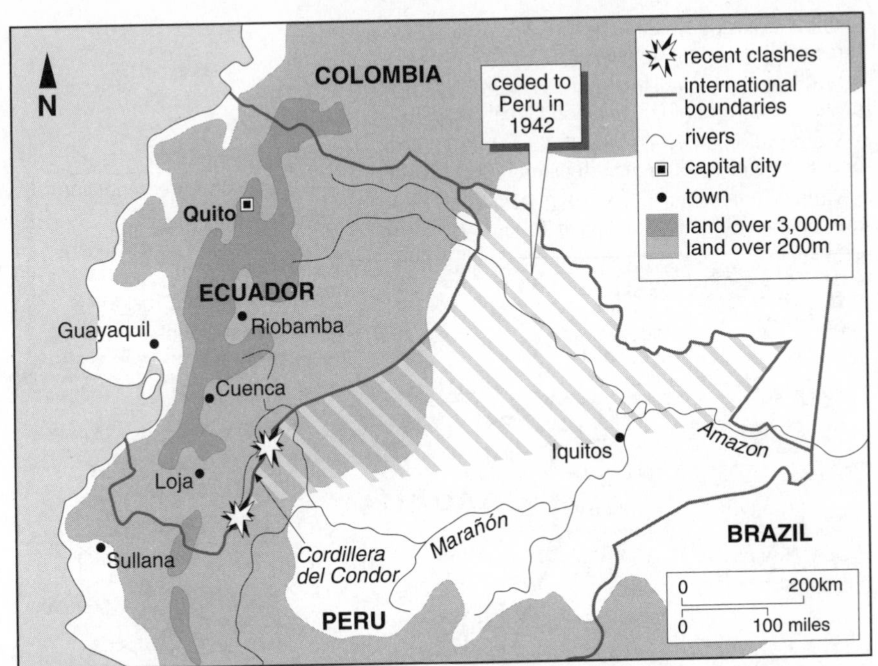

Military clashes between Peruvian and Ecuadorian patrols along the disputed frontier region of the Cordillera del Condor in January and February 1995 lasted for about five weeks. A cease-fire has since been monitored by the Military Observer Mission Ecuador/Peru (MOMEP). Tensions persist, however, with both countries conducting manoeuvres along the border in violation of cease-fire accords. In 1996, Peru also purchased 12 MiG-29 (*Fulcrum*) fighters and 14 Sukhoi-25 (*Frogfoot*) ground-attack aircraft from Belarus, ostensibly to replace losses from the 1995 border conflict.

SFOR consists of 31,000–32,000 troops divided into three multi-national divisions:

• Multi-national Division South West (MND/SE) is under the control of UK division HQ at Banja Luka

• Multi-national Division South East (MND/SE) is under the control of French division HQ at Mostar

• Multi-national Division North (MND/N) is under the control of US division HQ at Tuzla

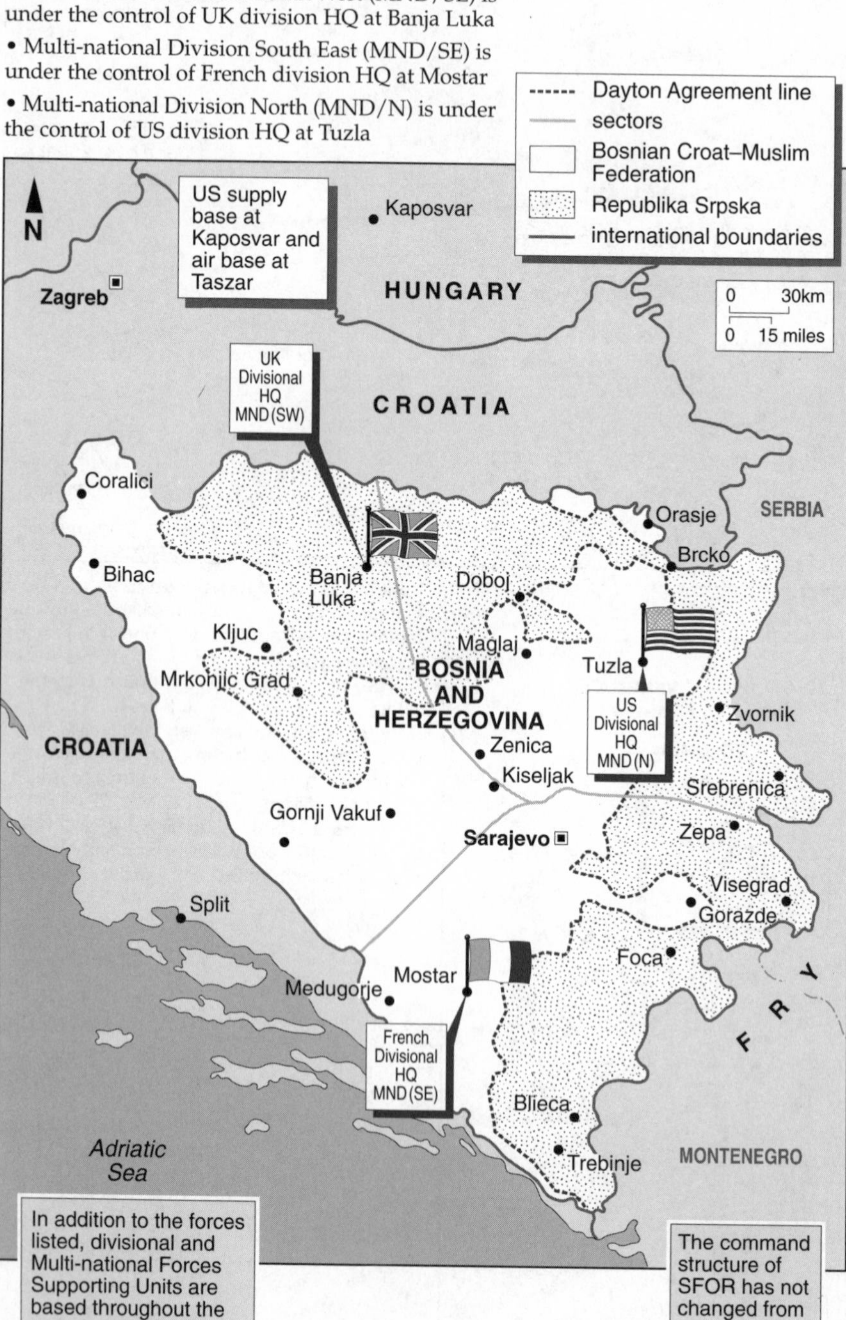

In addition to the forces listed, divisional and Multi-national Forces Supporting Units are based throughout the area and in Croatia

The command structure of SFOR has not changed from that of IFOR

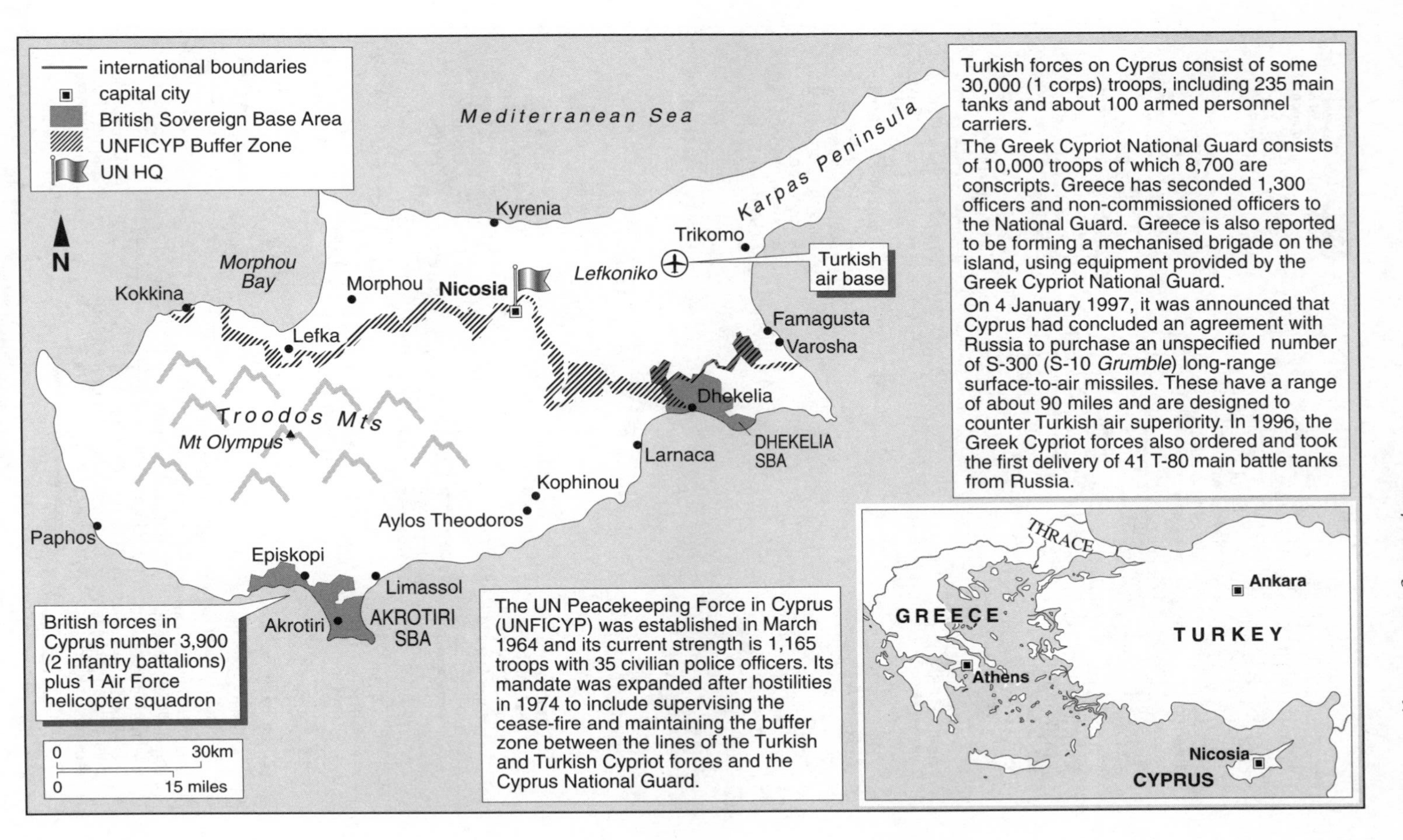
international boundaries
capital city
British Sovereign Base Area
UNFICYP Buffer Zone
UN HQ
N
Mediterranean Sea
Karpas Peninsula
Kyrenia
Trikomo
Lefkoniko
Turkish air base
Morphou Bay
Kokkina
Morphou
Nicosia
Lefka
Famagusta
Varosha
Dhekelia
DHEKELIA SBA
Troodos Mts
Mt Olympus
Larnaca
Kophinou
Aylos Theodoros
Paphos
Episkopi
Limassol
Akrotiri
AKROTIRI SBA
0 30km
0 15 miles
Turkish forces on Cyprus consist of some 30,000 (1 corps) troops, including 235 main tanks and about 100 armed personnel carriers.
The Greek Cypriot National Guard consists of 10,000 troops of which 8,700 are conscripts. Greece has seconded 1,300 officers and non-commissioned officers to the National Guard. Greece is also reported to be forming a mechanised brigade on the island, using equipment provided by the Greek Cypriot National Guard.
On 4 January 1997, it was announced that Cyprus had concluded an agreement with Russia to purchase an unspecified number of S-300 (S-10 Grumble) long-range surface-to-air missiles. These have a range of about 90 miles and are designed to counter Turkish air superiority. In 1996, the Greek Cypriot forces also ordered and took the first delivery of 41 T-80 main battle tanks from Russia.
British forces in Cyprus number 3,900 (2 infantry battalions) plus 1 Air Force helicopter squadron
The UN Peacekeeping Force in Cyprus (UNFICYP) was established in March 1964 and its current strength is 1,165 troops with 35 civilian police officers. Its mandate was expanded after hostilities in 1974 to include supervising the cease-fire and maintaining the buffer zone between the lines of the Turkish and Turkish Cypriot forces and the Cyprus National Guard.
THRACE
GREECE
Athens
TURKEY
Ankara
Nicosia
CYPRUS

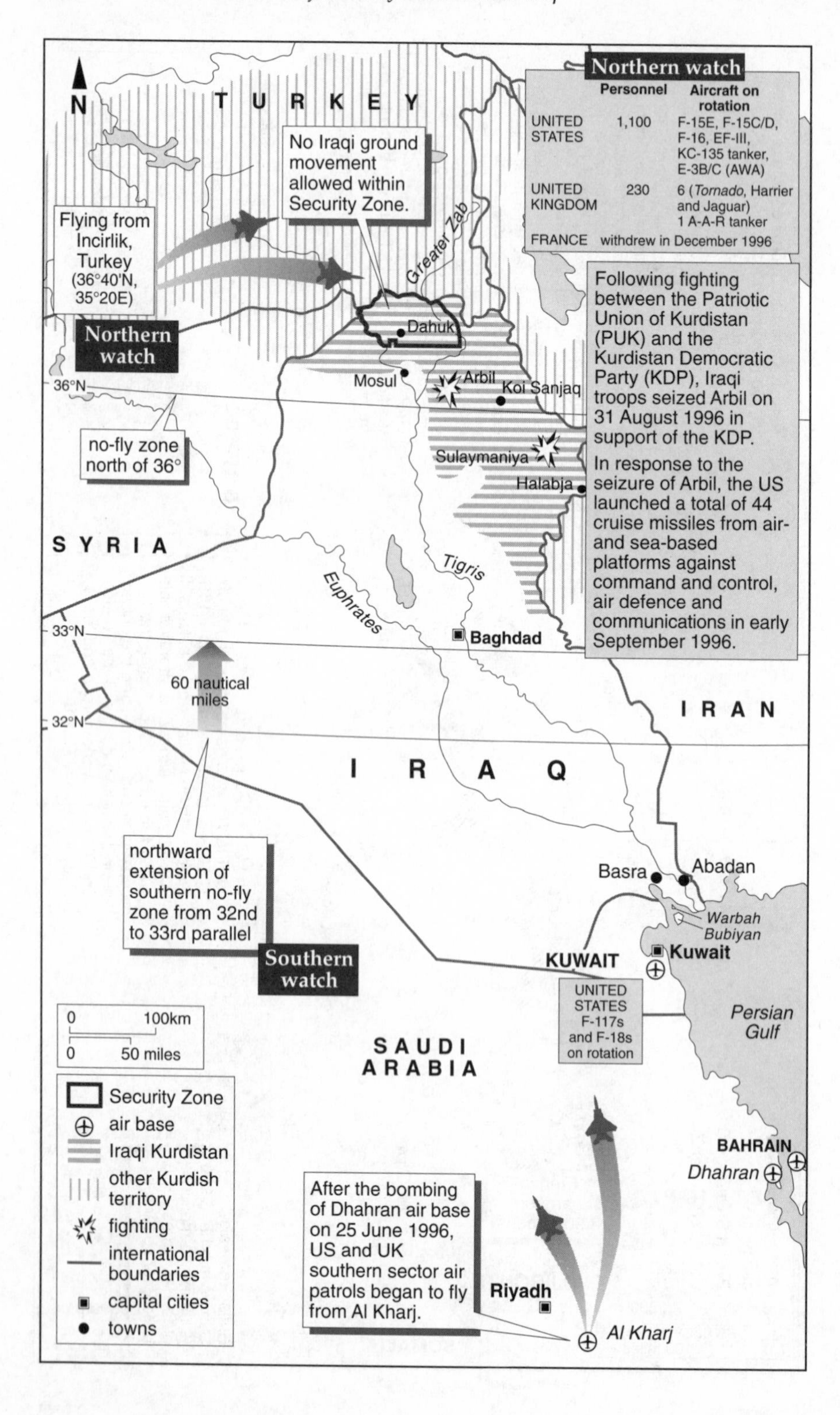

N
TURKEY
Northern watch
Personnel
Aircraft on rotation
UNITED STATES 1,100 F-15E, F-15C/D, F-16, EF-III, KC-135 tanker, E-3B/C (AWA)
UNITED KINGDOM 230 6 (Tornado, Harrier and Jaguar) 1 A-A-R tanker
FRANCE withdrew in December 1996
No Iraqi ground movement allowed within Security Zone.
Flying from Incirlik, Turkey (36°40'N, 35°20E)
Northern watch
Greater Zab
Following fighting between the Patriotic Union of Kurdistan (PUK) and the Kurdistan Democratic Party (KDP), Iraqi troops seized Arbil on 31 August 1996 in support of the KDP.
In response to the seizure of Arbil, the US launched a total of 44 cruise missiles from air- and sea-based platforms against command and control, air defence and communications in early September 1996.
Dahuk
Mosul
Arbil
Koi Sanjaq
Sulaymaniya
Halabja
36°N
no-fly zone north of 36°
SYRIA
Tigris
Euphrates
Baghdad
33°N
60 nautical miles
32°N
IRAN
IRAQ
northward extension of southern no-fly zone from 32nd to 33rd parallel
Southern watch
Basra
Abadan
Warbah
Bubiyan
KUWAIT
Kuwait
UNITED STATES F-117s and F-18s on rotation
Persian Gulf
0 100km
0 50 miles
SAUDI ARABIA
Security Zone
air base
Iraqi Kurdistan
other Kurdish territory
fighting
international boundaries
capital cities
towns
BAHRAIN
Dhahran
After the bombing of Dhahran air base on 25 June 1996, US and UK southern sector air patrols began to fly from Al Kharj.
Riyadh
Al Kharj

0 100km
0 50 miles
N
IRAN
Mosul
Euphrates
SYRIA
Samarra
Al Karama
Habbaniyah
Baghdad
Tuwaitha
Salman Pak
Al Musayyib
Latifyah
Al Hakam
Tigris
IRAQ
Basra
Abadan
KUWAIT
Kuwait
Persian Gulf
SAUDI ARABIA
ballistic-missile installation
biological installation
chemical installation
nuclear installation
international boundaries
capital cities
towns

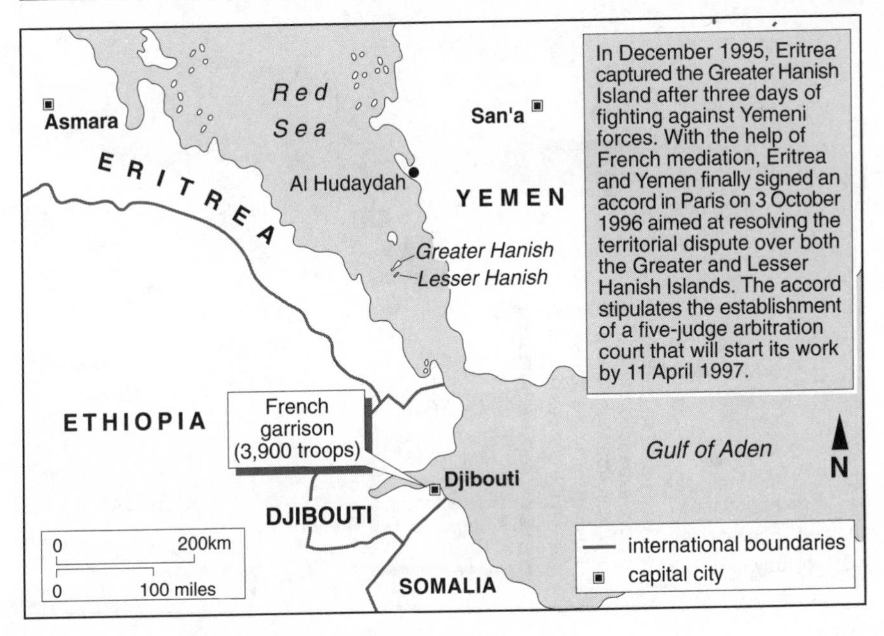
Asmara
Red Sea
San'a
ERITREA
Al Hudaydah
YEMEN
Greater Hanish
Lesser Hanish
In December 1995, Eritrea captured the Greater Hanish Island after three days of fighting against Yemeni forces. With the help of French mediation, Eritrea and Yemen finally signed an accord in Paris on 3 October 1996 aimed at resolving the territorial dispute over both the Greater and Lesser Hanish Islands. The accord stipulates the establishment of a five-judge arbitration court that will start its work by 11 April 1997.
ETHIOPIA
French garrison (3,900 troops)
Gulf of Aden
N
Djibouti
DJIBOUTI
0 200km
0 100 miles
SOMALIA
international boundaries
capital city

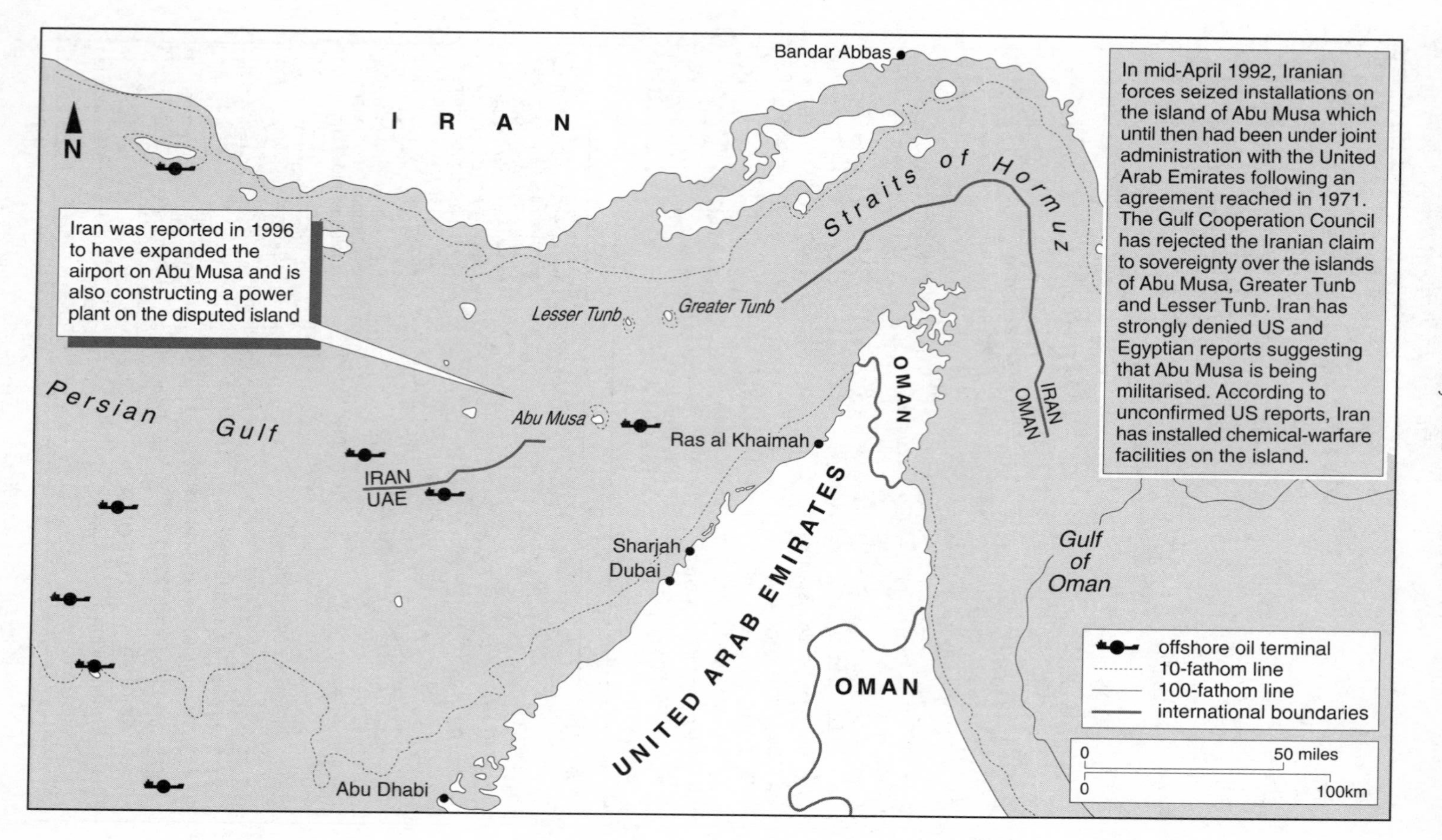
N
IRAN
Bandar Abbas
In mid-April 1992, Iranian forces seized installations on the island of Abu Musa which until then had been under joint administration with the United Arab Emirates following an agreement reached in 1971. The Gulf Cooperation Council has rejected the Iranian claim to sovereignty over the islands of Abu Musa, Greater Tunb and Lesser Tunb. Iran has strongly denied US and Egyptian reports suggesting that Abu Musa is being militarised. According to unconfirmed US reports, Iran has installed chemical-warfare facilities on the island.
Iran was reported in 1996 to have expanded the airport on Abu Musa and is also constructing a power plant on the disputed island
Straits of Hormuz
Lesser Tunb
Greater Tunb
Persian Gulf
Abu Musa
OMAN
IRAN
OMAN
Ras al Khaimah
IRAN
UAE
Sharjah
Dubai
UNITED ARAB EMIRATES
Gulf of Oman
OMAN
Abu Dhabi
offshore oil terminal
10-fathom line
100-fathom line
international boundaries
0
50 miles
0
100km

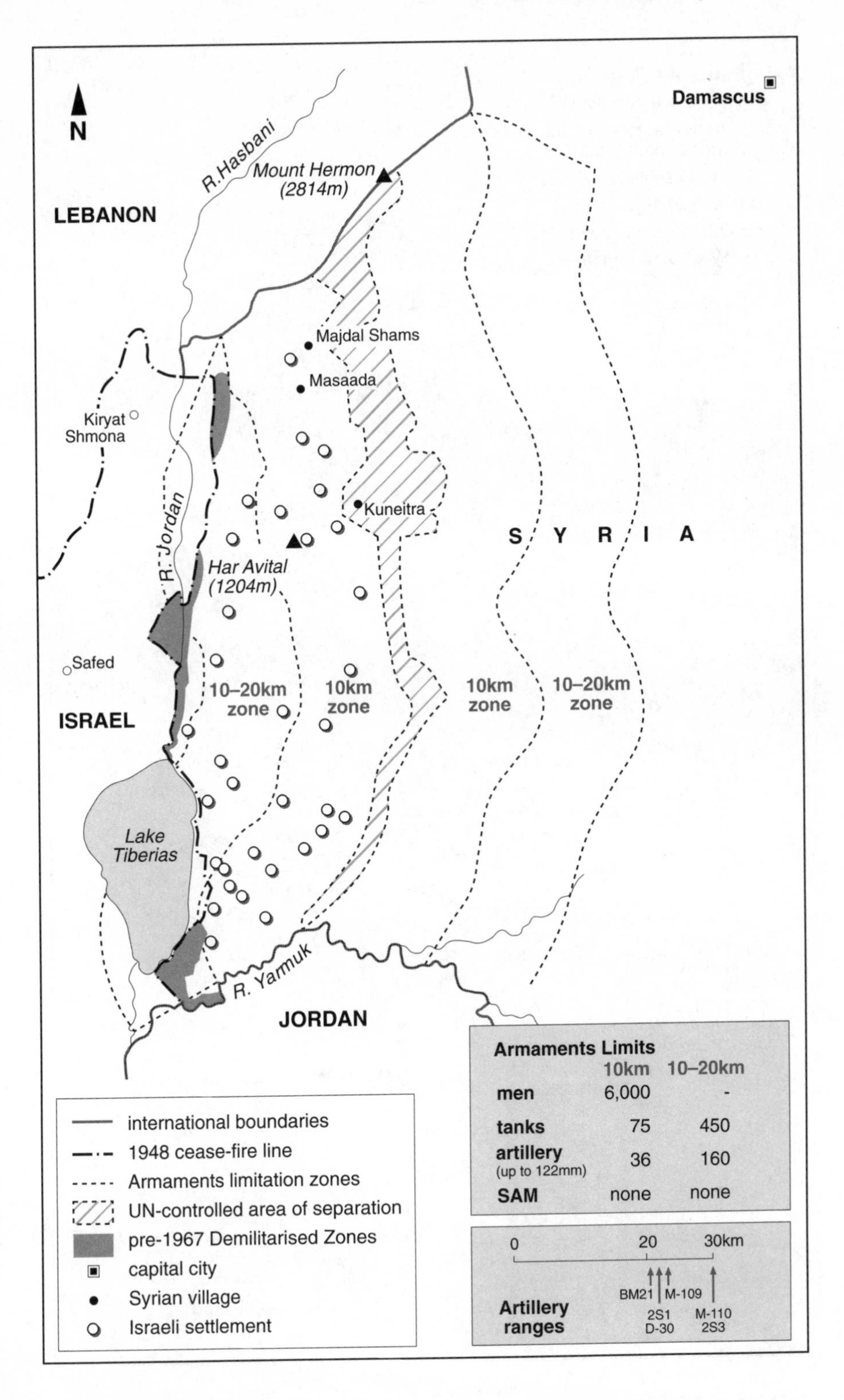

N
Damascus
LEBANON
R. Hasbani
Mount Hermon
(2814m)
Majdal Shams
Masaada
Kiryat Shmona
Kuneitra
R. Jordan
Har Avital
(1204m)
S Y R I A
Safed
ISRAEL
10–20km zone
10km zone
10km zone
10–20km zone
Lake Tiberias
R. Yarmuk
JORDAN
international boundaries
1948 cease-fire line
Armaments limitation zones
UN-controlled area of separation
pre-1967 Demilitarised Zones
capital city
Syrian village
Israeli settlement
Armaments Limits
10km 10–20km
men 6,000 -
tanks 75 450
artillery (up to 122mm) 36 160
SAM none none
0 20 30km
BM21 M-109
2S1 D-30
M-110 2S3
Artillery ranges

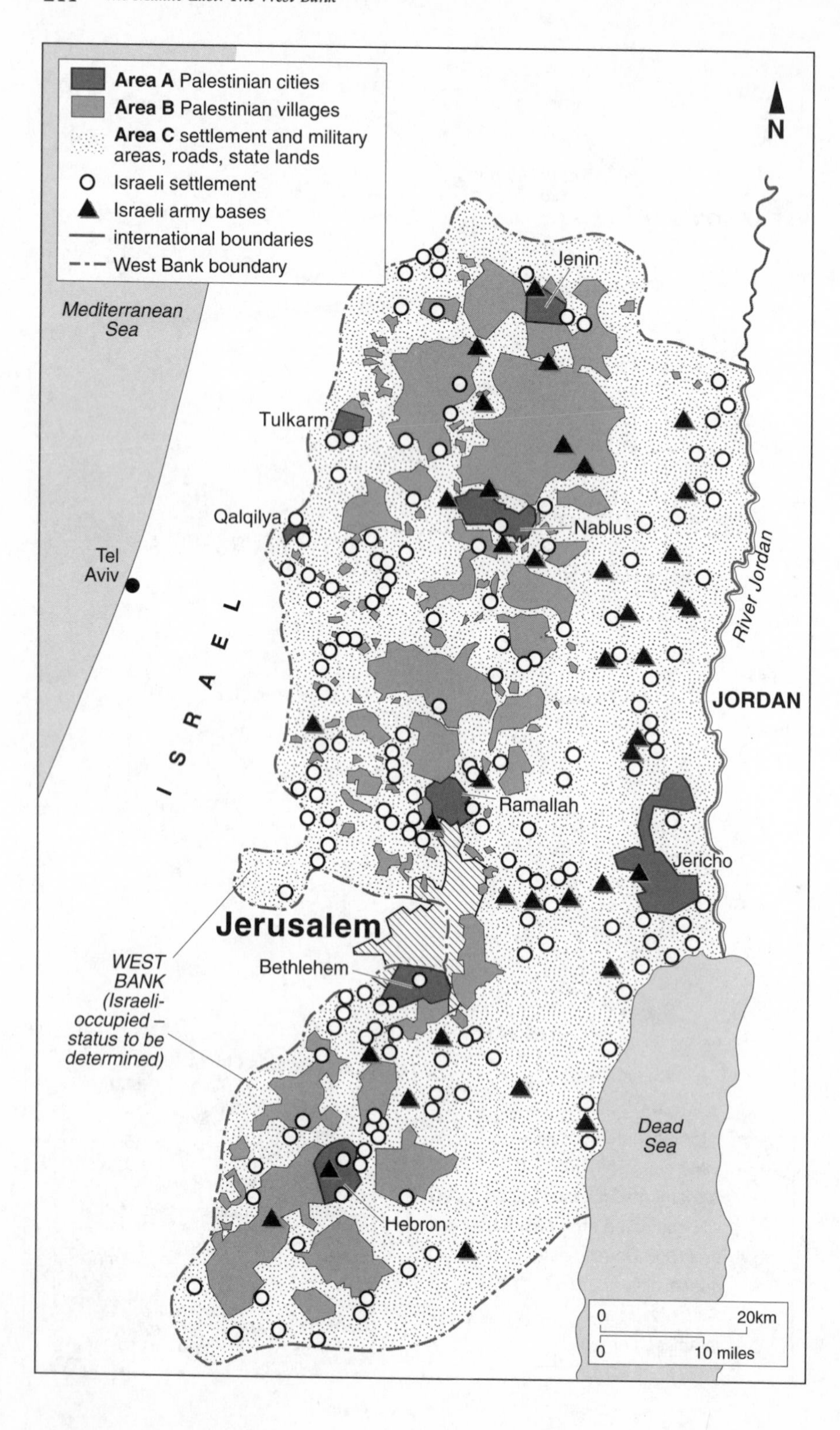
Area A Palestinian cities
Area B Palestinian villages
Area C settlement and military areas, roads, state lands
Israeli settlement
Israeli army bases
international boundaries
West Bank boundary
N
Mediterranean Sea
Tel Aviv
ISRAEL
Jenin
Tulkarm
Qalqilya
Nablus
River Jordan
JORDAN
Ramallah
Jericho
Jerusalem
Bethlehem
WEST BANK (Israeli-occupied – status to be determined)
Hebron
Dead Sea
0
20km
0
10 miles

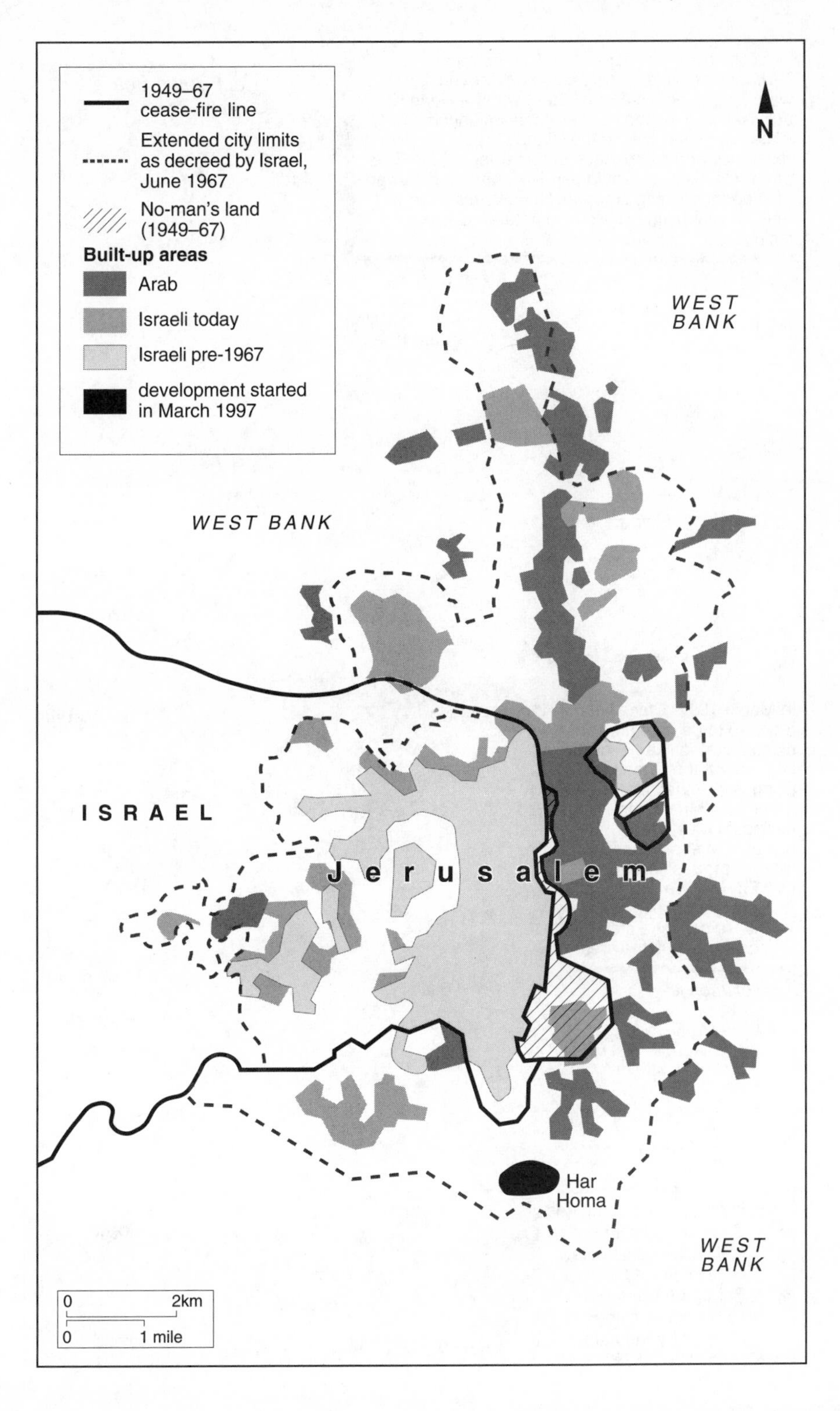
1949–67 cease-fire line
Extended city limits as decreed by Israel, June 1967
No-man's land (1949–67)
Built-up areas
Arab
Israeli today
Israeli pre-1967
development started in March 1997
N
WEST BANK
WEST BANK
ISRAEL
Jerusalem
Har Homa
WEST BANK
0 2km
0 1 mile

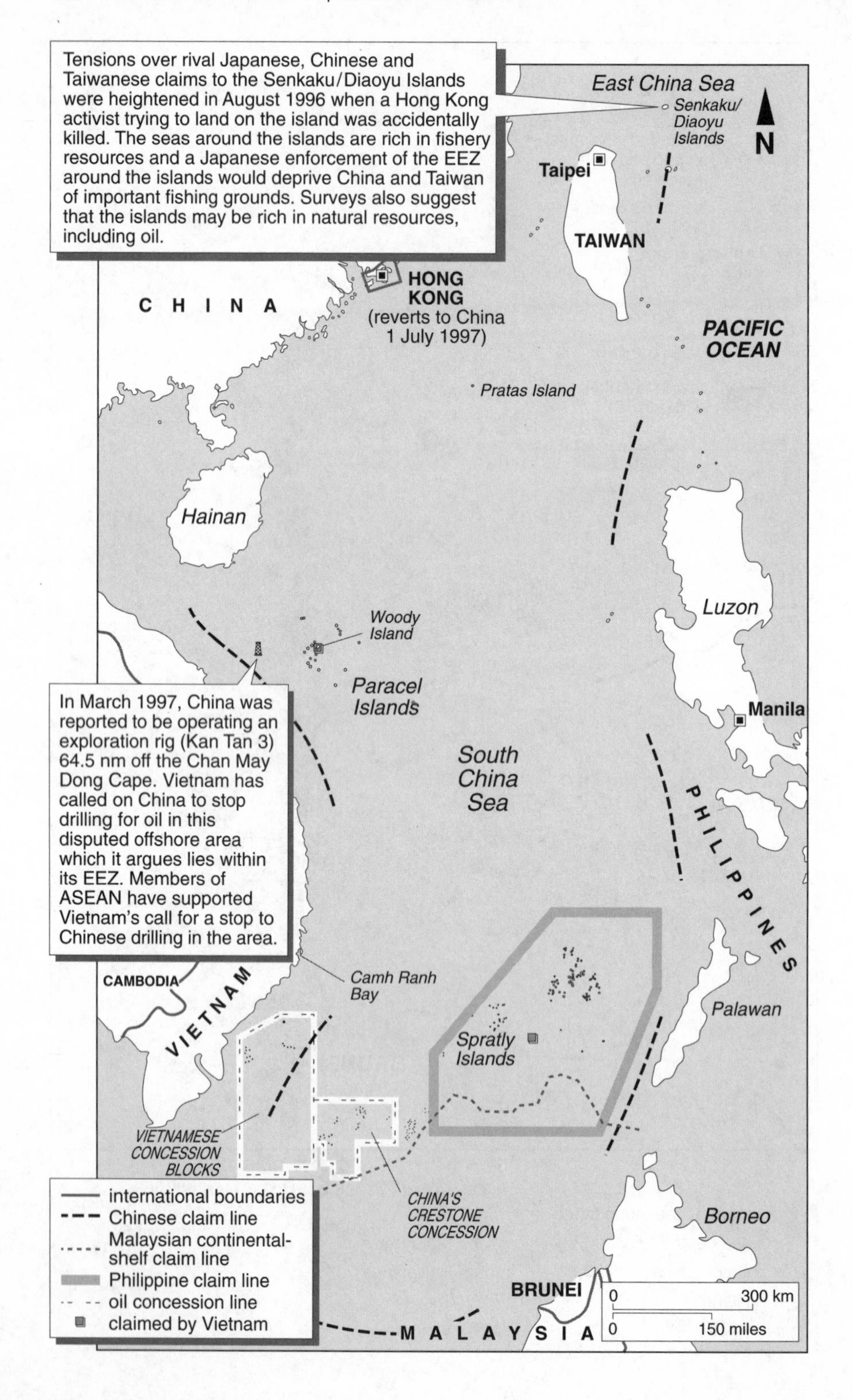
Tensions over rival Japanese, Chinese and Taiwanese claims to the Senkaku/Diaoyu Islands were heightened in August 1996 when a Hong Kong activist trying to land on the island was accidentally killed. The seas around the islands are rich in fishery resources and a Japanese enforcement of the EEZ around the islands would deprive China and Taiwan of important fishing grounds. Surveys also suggest that the islands may be rich in natural resources, including oil.
East China Sea
Senkaku/ Diaoyu Islands
N
Taipei
TAIWAN
HONG KONG (reverts to China 1 July 1997)
C H I N A
PACIFIC OCEAN
Pratas Island
Hainan
Luzon
Woody Island
Paracel Islands
Manila
South China Sea
PHILIPPINES
In March 1997, China was reported to be operating an exploration rig (Kan Tan 3) 64.5 nm off the Chan May Dong Cape. Vietnam has called on China to stop drilling for oil in this disputed offshore area which it argues lies within its EEZ. Members of ASEAN have supported Vietnam's call for a stop to Chinese drilling in the area.
CAMBODIA
VIETNAM
Camh Ranh Bay
Palawan
Spratly Islands
VIETNAMESE CONCESSION BLOCKS
CHINA'S CRESTONE CONCESSION
Borneo
international boundaries
Chinese claim line
Malaysian continental-shelf claim line
Philippine claim line
oil concession line
claimed by Vietnam
BRUNEI
MALAYSIA
0 300 km
0 150 miles

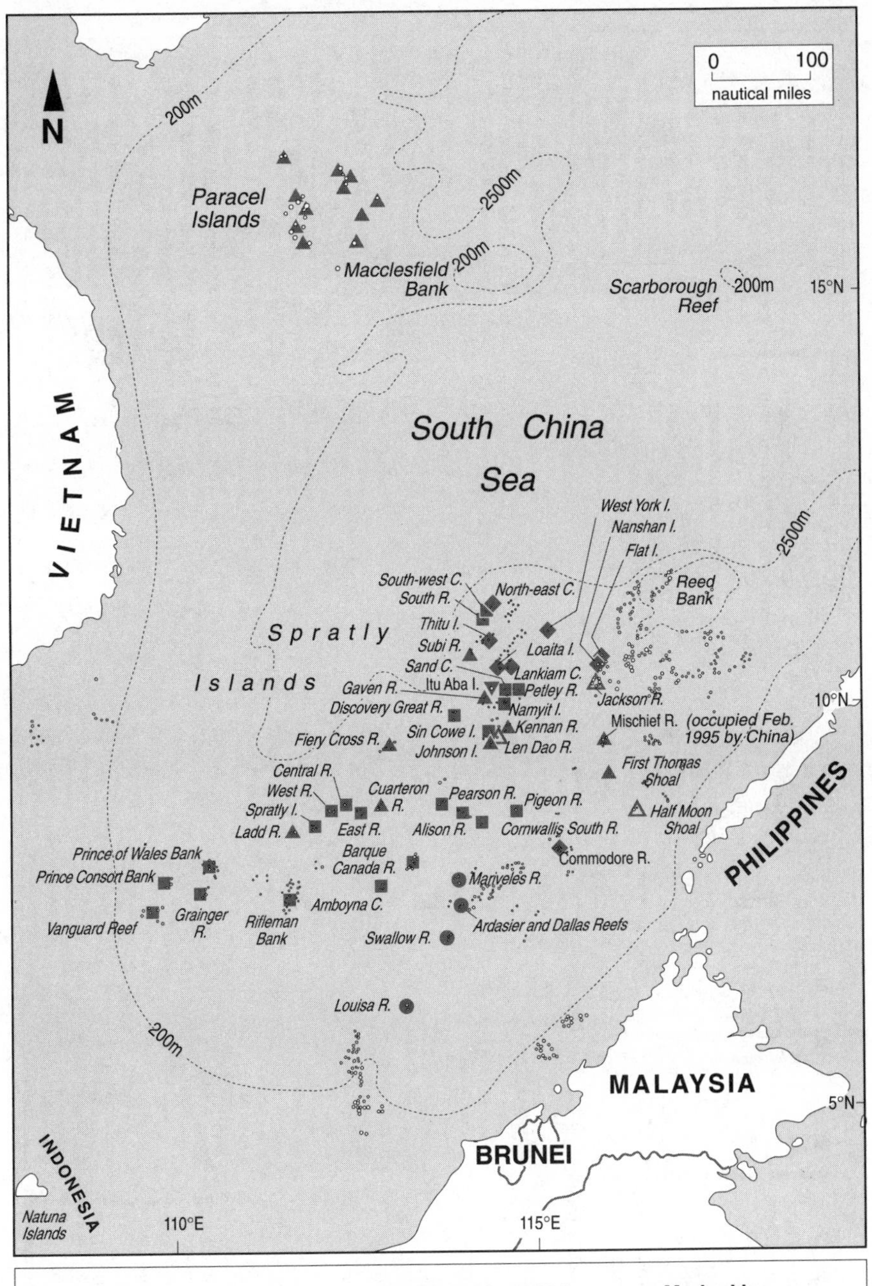
0 100
nautical miles
N
200m
2500m
Paracel
Islands
Macclesfield
Bank
Scarborough
Reef
15°N
VIETNAM
South China
Sea
West York I.
Nanshan I.
Flat I.
South-west C.
North-east C.
Reed
Bank
South R.
Thitu I.
Spratly
Islands
Subi R.
Loaita I.
Sand C.
Lankiam C.
Gaven R.
Itu Aba I.
Petley R.
Discovery Great R.
Namyit I.
Jackson R.
10°N
Mischief R. (occupied Feb. 1995 by China)
Sin Cowe I.
Kennan R.
Fiery Cross R.
Johnson I.
Len Dao R.
First Thomas
Shoal
Central R.
West R.
Cuarteron
R.
Pearson R.
Pigeon R.
Spratly I.
Half Moon
Shoal
Ladd R.
East R.
Alison R.
Cornwallis South R.
PHILIPPINES
Prince of Wales Bank
Barque
Canada R.
Commodore R.
Prince Consort Bank
Mariveles R.
Amboyna C.
Grainger
R.
Vanguard Reef
Rifleman
Bank
Ardasier and Dallas Reefs
Swallow R.
Louisa R.
MALAYSIA
5°N
BRUNEI
INDONESIA
Natuna
Islands
110°E
115°E

international boundaries
sea depths
C. Cay
I. Island
R. Reef
Occupied in 1995 by:
China
Malaysia
Philippines
Taiwan
Vietnam
Marked by:
China (but
not occupied)

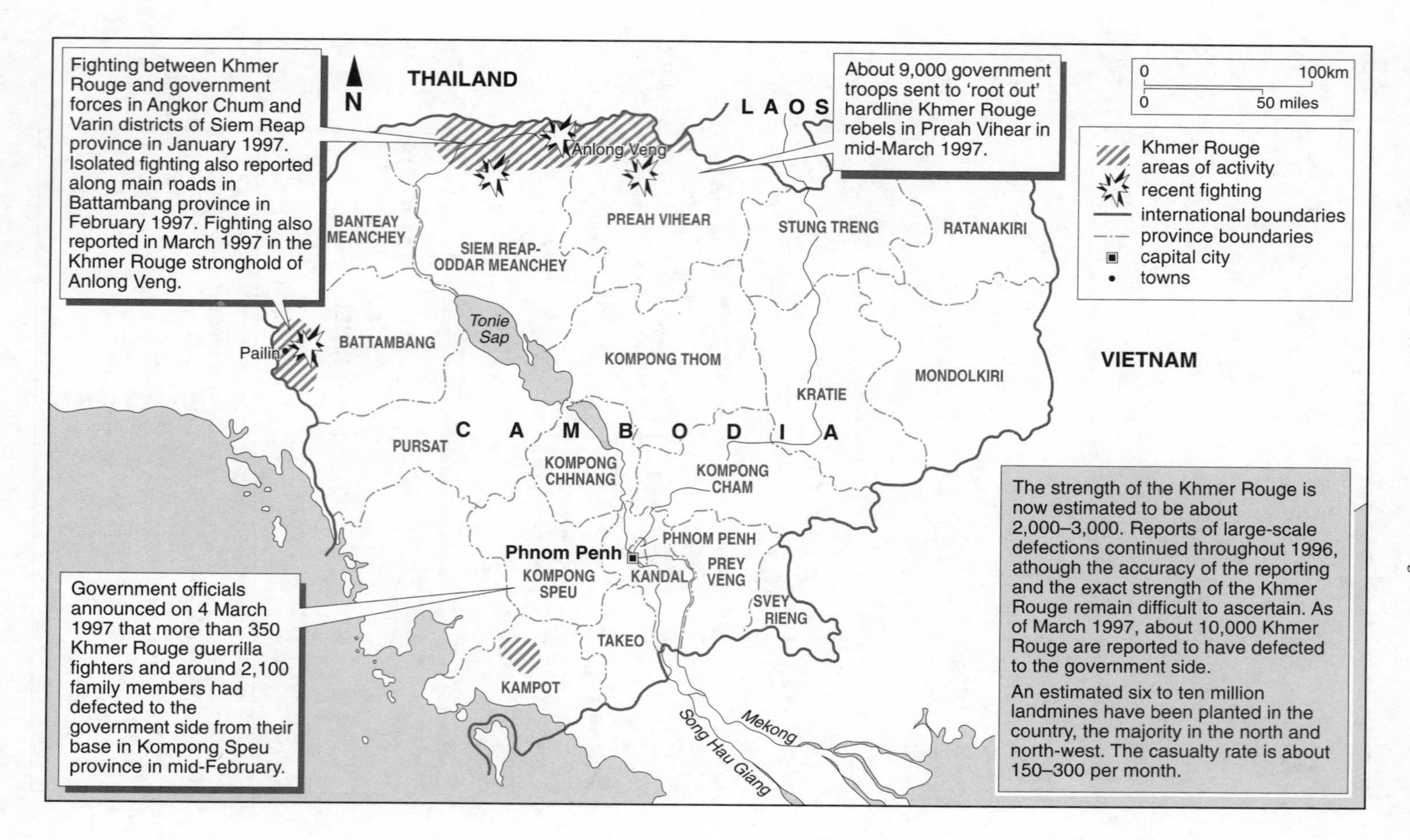
Fighting between Khmer Rouge and government forces in Angkor Chum and Varin districts of Siem Reap province in January 1997. Isolated fighting also reported along main roads in Battambang province in February 1997. Fighting also reported in March 1997 in the Khmer Rouge stronghold of Anlong Veng.
N
THAILAND
LAOS
About 9,000 government troops sent to 'root out' hardline Khmer Rouge rebels in Preah Vihear in mid-March 1997.
0
100km
0
50 miles
Khmer Rouge areas of activity
recent fighting
international boundaries
province boundaries
capital city
towns
Anlong Veng
BANTEAY MEANCHEY
SIEM REAP-ODDAR MEANCHEY
PREAH VIHEAR
STUNG TRENG
RATANAKIRI
Pailin
BATTAMBANG
Tonie Sap
KOMPONG THOM
MONDOLKIRI
KRATIE
VIETNAM
CAMBODIA
PURSAT
KOMPONG CHHNANG
KOMPONG CHAM
Phnom Penh
PHNOM PENH
KOMPONG SPEU
KANDAL
PREY VENG
SVEY RIENG
TAKEO
KAMPOT
Song Hau Giang
Mekong
Government officials announced on 4 March 1997 that more than 350 Khmer Rouge guerrilla fighters and around 2,100 family members had defected to the government side from their base in Kompong Speu province in mid-February.
The strength of the Khmer Rouge is now estimated to be about 2,000–3,000. Reports of large-scale defections continued throughout 1996, athough the accuracy of the reporting and the exact strength of the Khmer Rouge remain difficult to ascertain. As of March 1997, about 10,000 Khmer Rouge are reported to have defected to the government side.
An estimated six to ten million landmines have been planted in the country, the majority in the north and north-west. The casualty rate is about 150–300 per month.

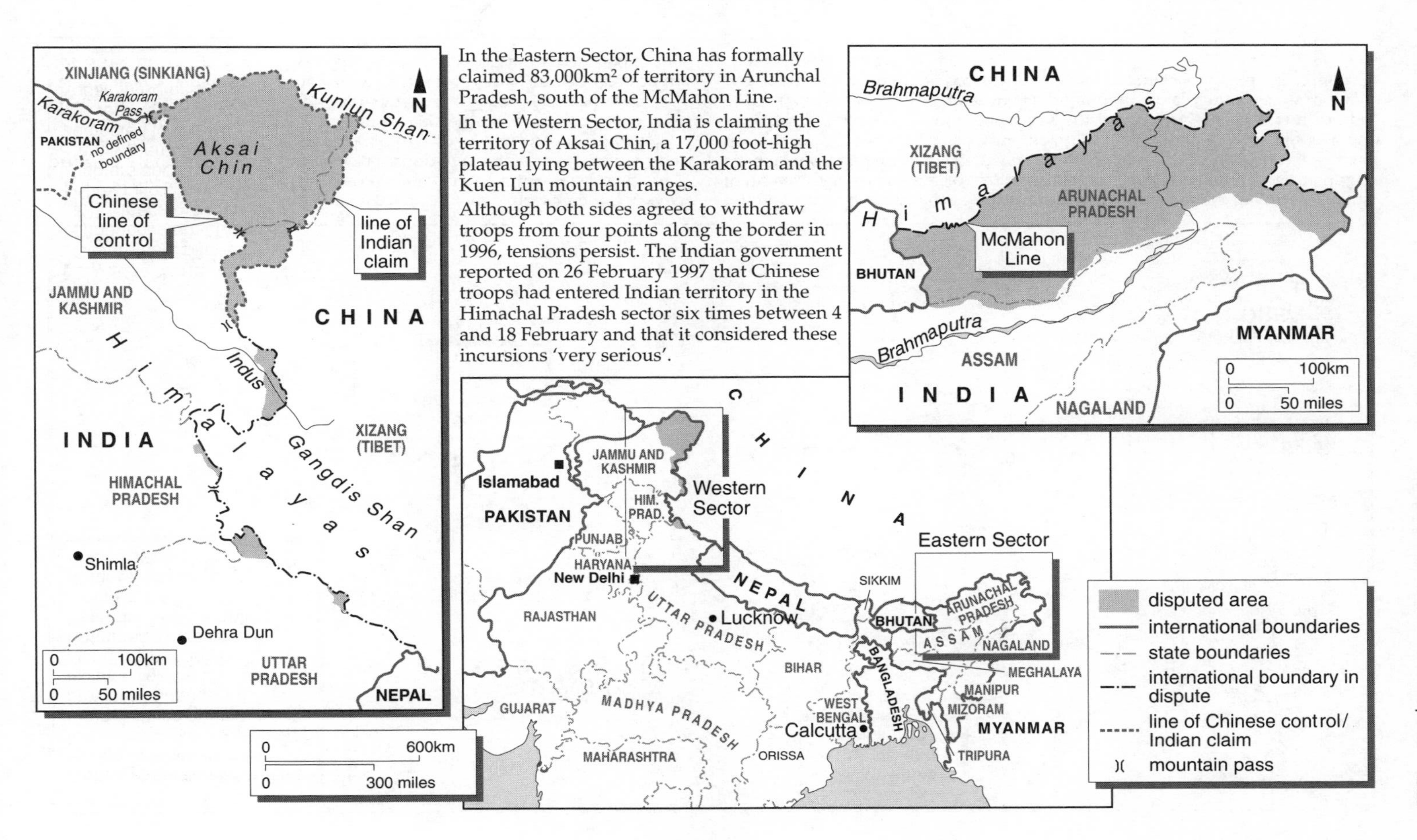

In the Eastern Sector, China has formally claimed 83,000km² of territory in Arunchal Pradesh, south of the McMahon Line.

In the Western Sector, India is claiming the territory of Aksai Chin, a 17,000 foot-high plateau lying between the Karakoram and the Kuen Lun mountain ranges.

Although both sides agreed to withdraw troops from four points along the border in 1996, tensions persist. The Indian government reported on 26 February 1997 that Chinese troops had entered Indian territory in the Himachal Pradesh sector six times between 4 and 18 February and that it considered these incursions 'very serious'.

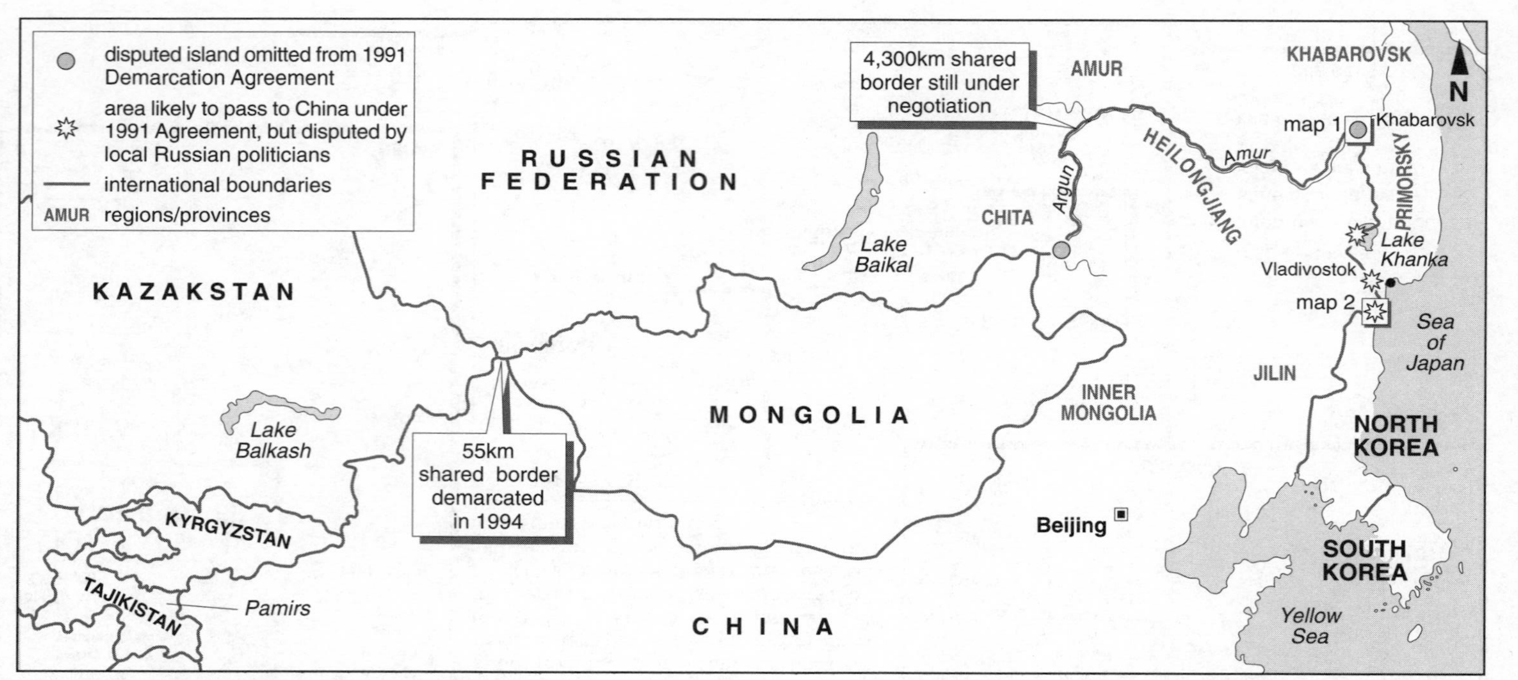

China and Russia share two borders – a 55km section to the west and a 4,300km stretch to the east. In 1991, China and the former Soviet Union agreed on principles for demarcating the eastern section of the disputed border as part of normalising relations. Russia pledged to respect that agreement in 1992. Annual negotiations have been held since, concentrating on two broad areas. First, those omitted from the 1991 agreement: the 55km Western Sector, two islands near Khabarovsk on the Amur River (see Map 1) and another on the upper reaches of the Argun River. The Western Sector was demarcated in 1994, but no principles have been outlined to resolve control over the three islands to the east. The second area of negotiation is demarcation of the eastern border. Demarcation under the 1991 agreement, athough never publicly outlined, has been fiercely resisted by local leaders within the Russian Far East. Revisions have been called for in a number of key areas including a 300 hectare section of the 150m-wide Tumen River which potentially gives China access to the Sea of Japan (see Map 2). According to Russian officials, final demarcation should be completed by the end of 1997.

China also concluded an agreement with Kyrgyzstan demarcating disputed sections of their border in 1994. China has outstanding claims to the Pamirs in Tajikistan, but is not pressing to reach an early settlement.

In April 1996, China, Russia and the three Central Asian states bordering China – Kazakstan, Kyrgyzstan and Tajikistan – reached an *Agreement on Confidence-Building in the Military Field along the Border Areas*. A limited demilitarisation agreement covering 100km either side of their borders is due to be signed in April 1997 at the presidential summit in Moscow.

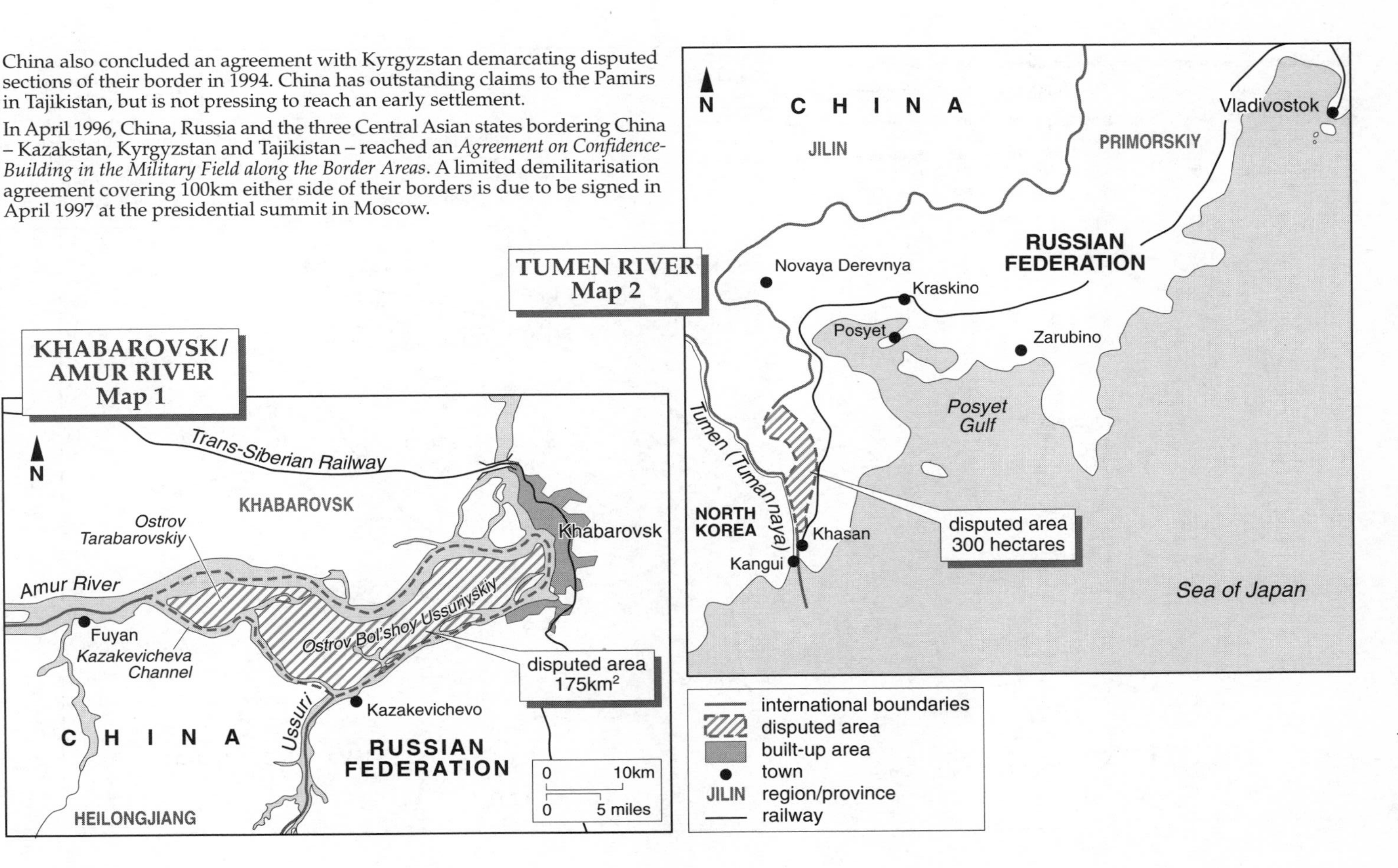

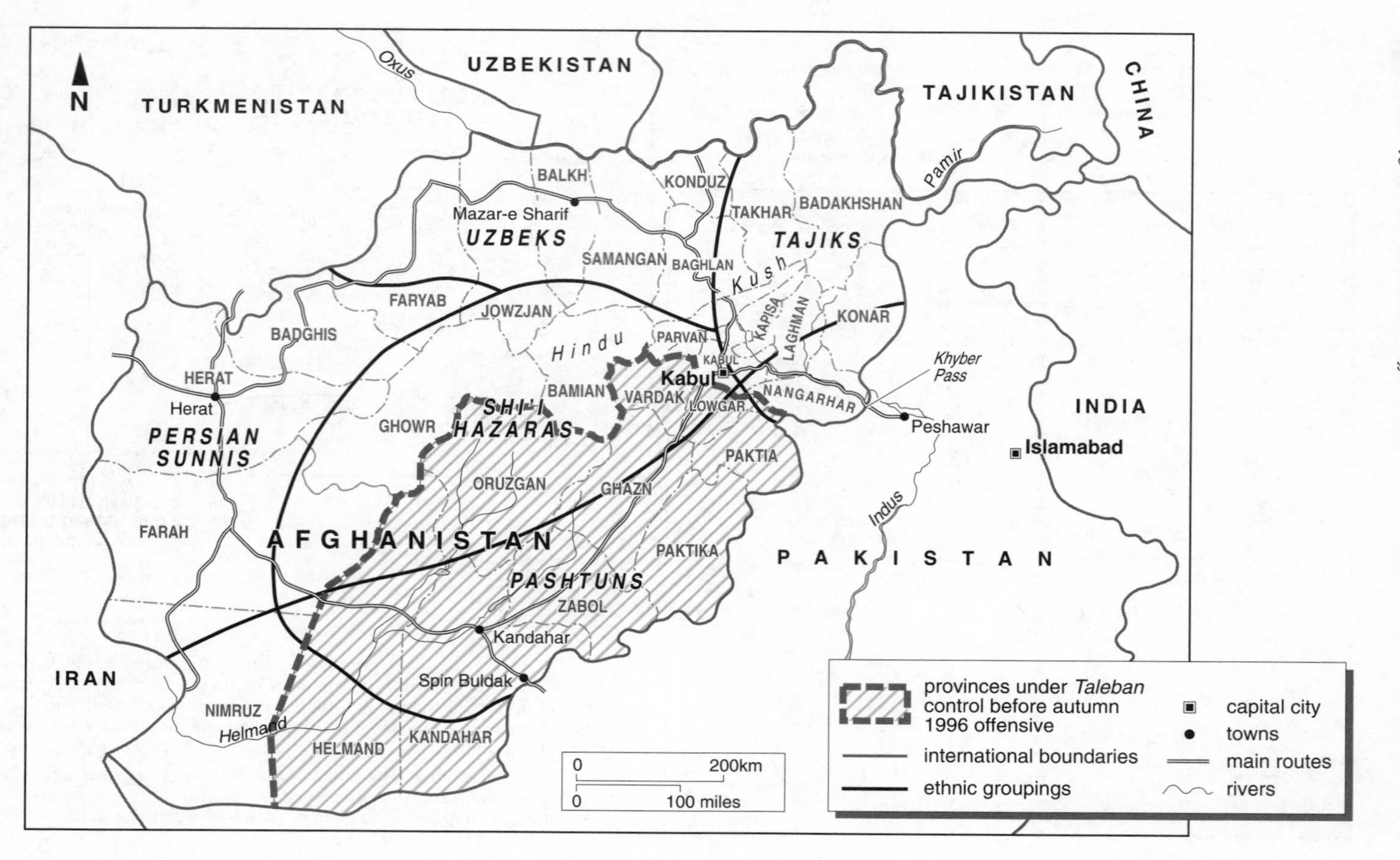
N
TURKMENISTAN
UZBEKISTAN
TAJIKISTAN
CHINA
INDIA
P A K I S T A N
IRAN
A F G H A N I S T A N
Oxus
Pamir
Indus
Helmand
Khyber Pass
Islamabad
Peshawar
Kabul
Mazar-e Sharif
Herat
Kandahar
Spin Buldak
Hindu
Kush
UZBEKS
TAJIKS
SHI'I HAZARAS
PERSIAN SUNNIS
PASHTUNS
BALKH
KONDUZ
TAKHAR
BADAKHSHAN
SAMANGAN
BAGHLAN
FARYAB
JOWZJAN
BADGHIS
PARVAN
KAPISA
LAGHMAN
KONAR
KABUL
HERAT
BAMIAN
VARDAK
LOWGAR
NANGARHAR
GHOWR
PAKTIA
ORUZGAN
GHAZN
FARAH
PAKTIKA
ZABOL
NIMRUZ
HELMAND
KANDAHAR
0 200km
0 100 miles
provinces under *Taleban* control before autumn 1996 offensive
international boundaries
ethnic groupings
capital city
towns
main routes
rivers

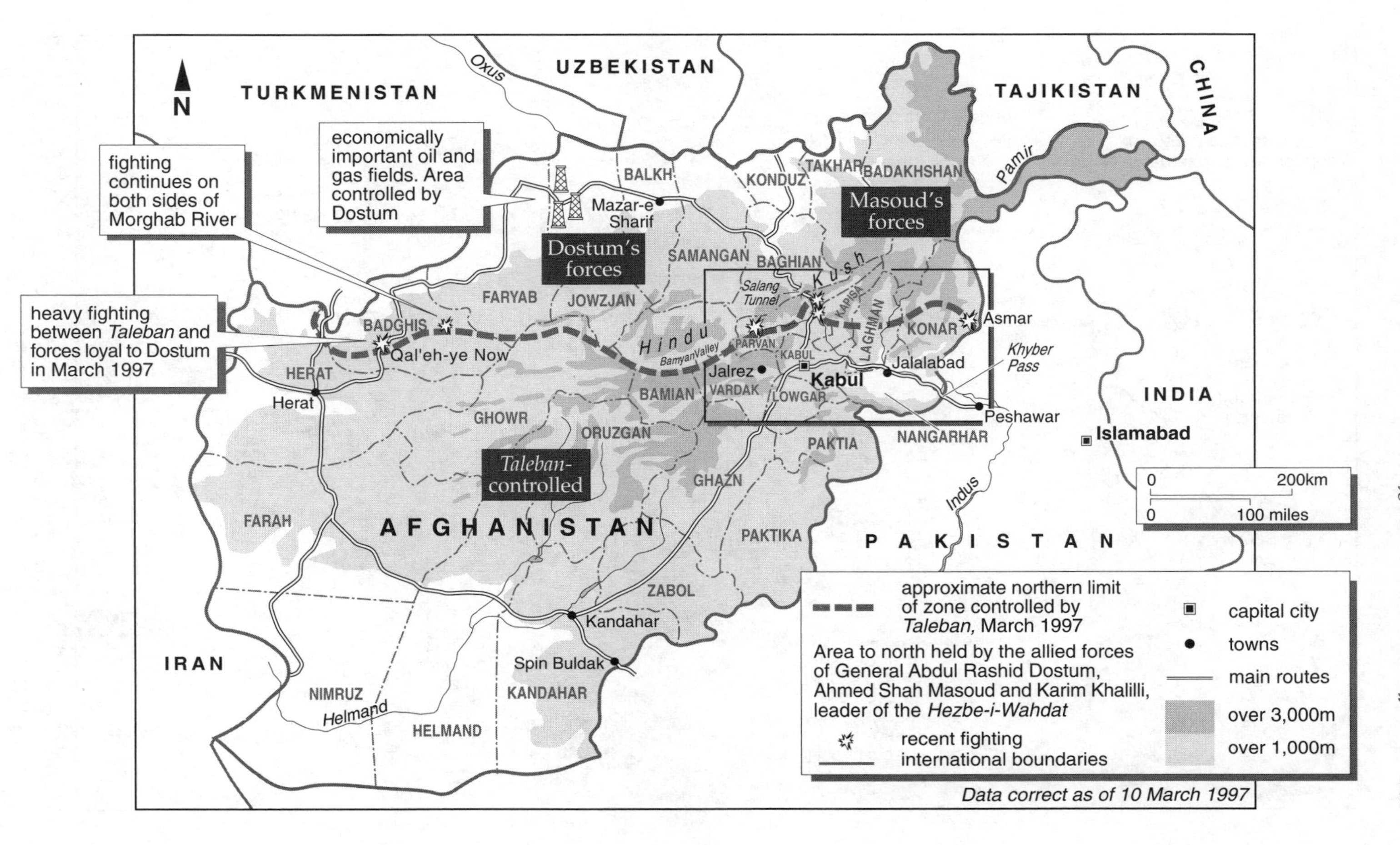

N
TURKMENISTAN
UZBEKISTAN
TAJIKISTAN
CHINA
INDIA
PAKISTAN
IRAN
AFGHANISTAN
Oxus
Pamir
Indus
Helmand
economically important oil and gas fields. Area controlled by Dostum
fighting continues on both sides of Morghab River
heavy fighting between *Taleban* and forces loyal to Dostum in March 1997
Dostum's forces
Masoud's forces
Taleban-controlled
BALKH
KONDUZ
TAKHAR
BADAKHSHAN
SAMANGAN
BAGHIAN
FARYAB
JOWZJAN
BADGHIS
HERAT
GHOWR
BAMIAN
ORUZGAN
PARVAN
KABUL
KAPISA
LAGHMAN
KONAR
VARDAK
LOWGAR
NANGARHAR
PAKTIA
GHAZNI
PAKTIKA
ZABOL
FARAH
NIMRUZ
KANDAHAR
HELMAND
Mazar-e Sharif
Qal'eh-ye Now
Herat
Salang Tunnel
Hindu Kush
Bamyan Valley
Jalrez
Kabul
Jalalabad
Asmar
Khyber Pass
Peshawar
Islamabad
Kandahar
Spin Buldak
0 200km
0 100 miles
approximate northern limit of zone controlled by *Taleban,* March 1997
Area to north held by the allied forces of General Abdul Rashid Dostum, Ahmed Shah Masoud and Karim Khalilli, leader of the *Hezbe-i-Wahdat*
recent fighting
international boundaries
capital city
towns
main routes
over 3,000m
over 1,000m
Data correct as of 10 March 1997

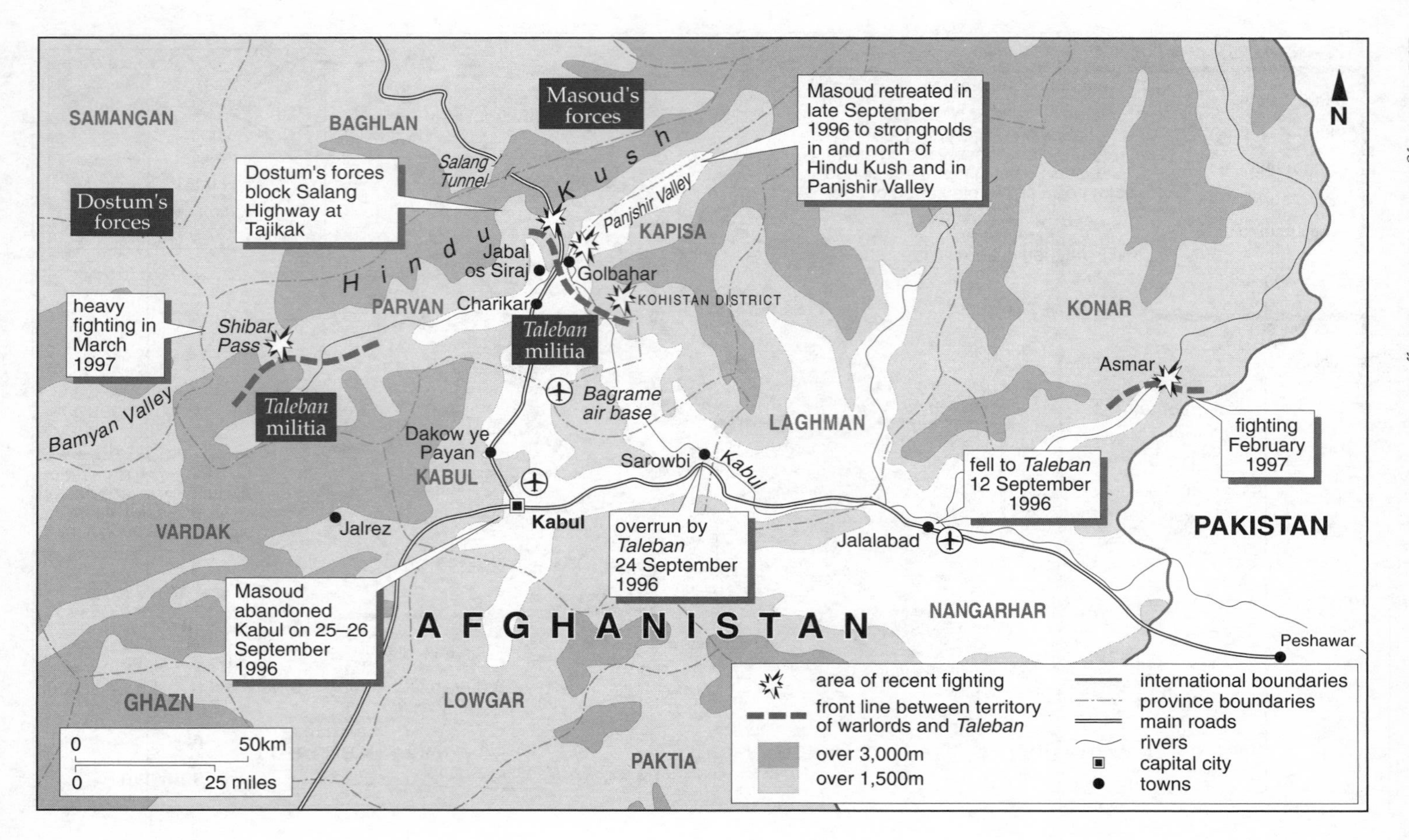
N
SAMANGAN
BAGHLAN
Masoud's forces
Masoud retreated in late September 1996 to strongholds in and north of Hindu Kush and in Panjshir Valley
Dostum's forces
Salang Tunnel
Dostum's forces block Salang Highway at Tajikak
Hindu Kush
Panjshir Valley
KAPISA
Jabal os Siraj
Golbahar
KOHISTAN DISTRICT
heavy fighting in March 1997
PARVAN
Charikar
KONAR
Shibar Pass
Taleban militia
Asmar
Bagrame air base
Taleban militia
LAGHMAN
Bamyan Valley
Dakow ye Payan
fighting February 1997
Sarowbi
Kabul
fell to *Taleban* 12 September 1996
KABUL
VARDAK
Jalrez
Kabul
overrun by *Taleban* 24 September 1996
Jalalabad
PAKISTAN
Masoud abandoned Kabul on 25–26 September 1996
AFGHANISTAN
NANGARHAR
Peshawar
GHAZN
LOWGAR
PAKTIA
0 50km
0 25 miles
area of recent fighting
front line between territory of warlords and *Taleban*
over 3,000m
over 1,500m
international boundaries
province boundaries
main roads
rivers
capital city
towns

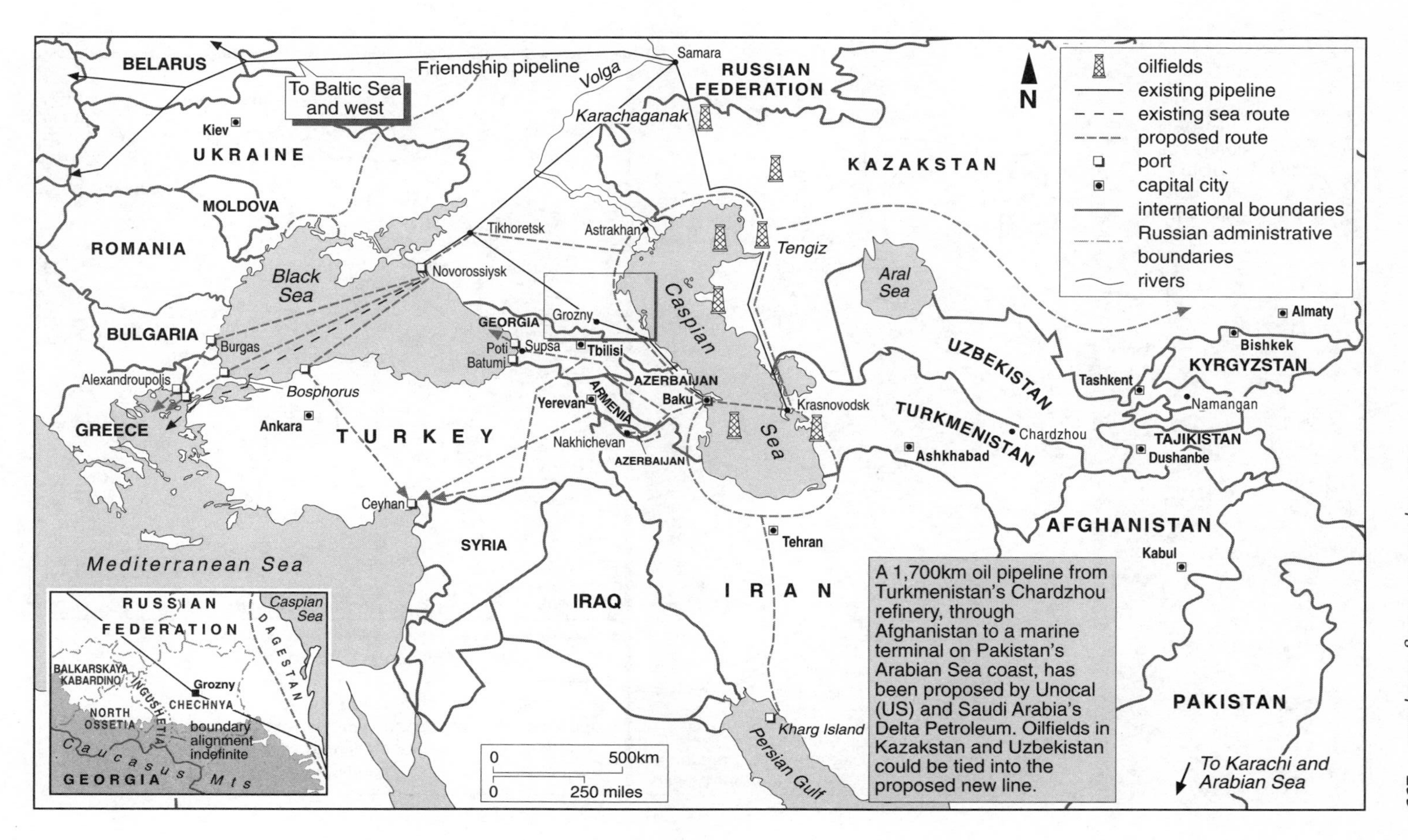
oilfields
existing pipeline
existing sea route
proposed route
port
capital city
international boundaries
Russian administrative boundaries
rivers
N
BELARUS
To Baltic Sea and west
Friendship pipeline
Volga
Samara
RUSSIAN FEDERATION
Karachaganak
KAZAKSTAN
Kiev
UKRAINE
MOLDOVA
ROMANIA
Tikhoretsk
Astrakhan
Tengiz
Aral Sea
Black Sea
Novorossiysk
Caspian Sea
Almaty
BULGARIA
Burgas
GEORGIA
Grozny
Poti
Supsa
Batumi
Tbilisi
Bishkek
UZBEKISTAN
KYRGYZSTAN
Alexandroupolis
Bosphorus
AZERBAIJAN
ARMENIA
Yerevan
Baku
Krasnovodsk
Tashkent
Namangan
TURKMENISTAN
Chardzhou
GREECE
Ankara
TURKEY
Nakhichevan
AZERBAIJAN
Ashkhabad
TAJIKISTAN
Dushanbe
Ceyhan
AFGHANISTAN
SYRIA
Tehran
Kabul
Mediterranean Sea
IRAQ
IRAN
RUSSIAN FEDERATION
Caspian Sea
DAGESTAN
BALKARSKAYA KABARDINO
INGUSHETIA
Grozny
CHECHNYA
NORTH OSSETIA
boundary alignment indefinite
Caucasus Mts
GEORGIA
PAKISTAN
Kharg Island
Persian Gulf
0 500km
0 250 miles
A 1,700km oil pipeline from Turkmenistan's Chardzhou refinery, through Afghanistan to a marine terminal on Pakistan's Arabian Sea coast, has been proposed by Unocal (US) and Saudi Arabia's Delta Petroleum. Oilfields in Kazakstan and Uzbekistan could be tied into the proposed new line.
To Karachi and Arabian Sea

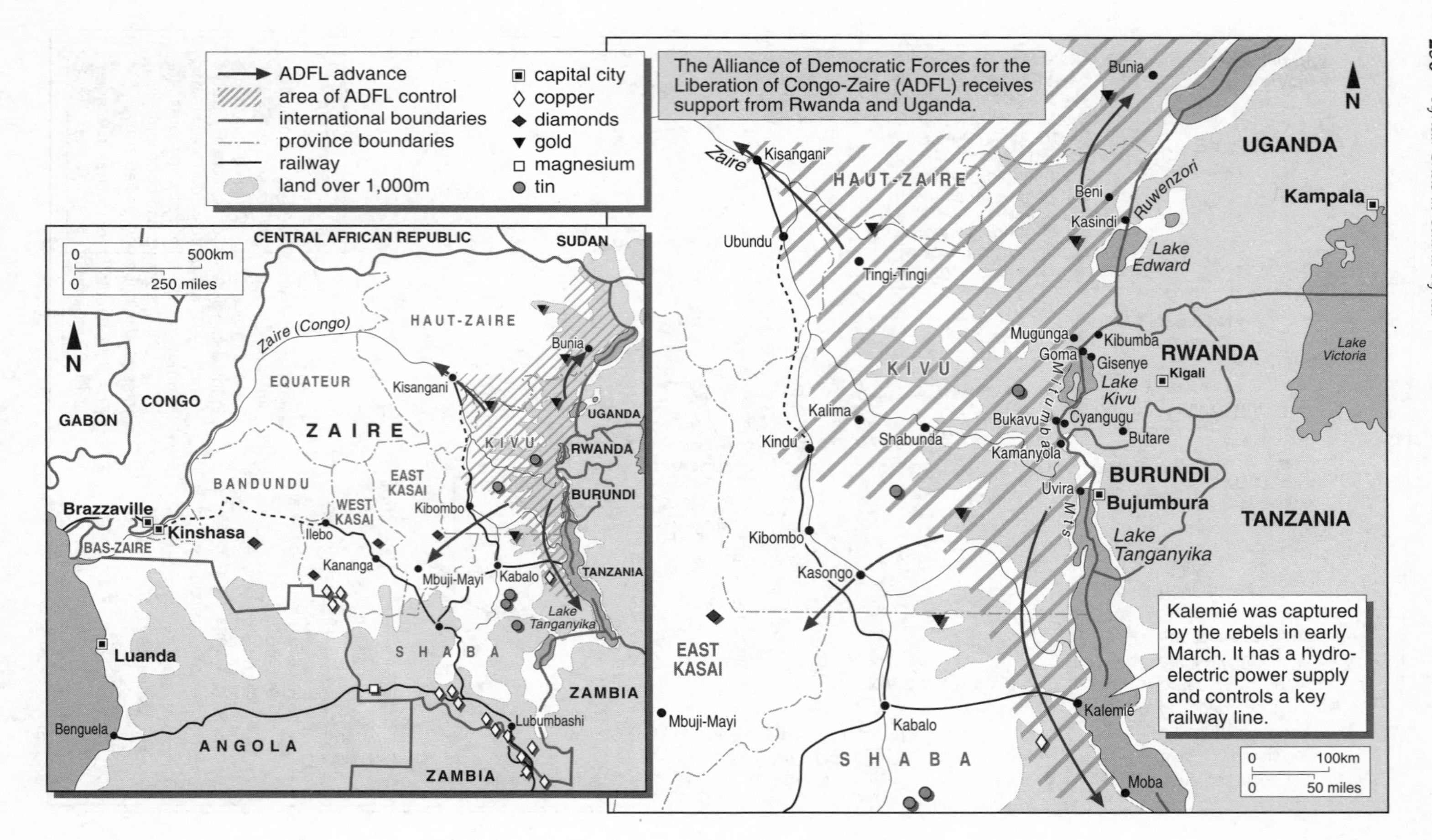
ADFL advance
area of ADFL control
international boundaries
province boundaries
railway
land over 1,000m
capital city
copper
diamonds
gold
magnesium
tin
The Alliance of Democratic Forces for the Liberation of Congo-Zaire (ADFL) receives support from Rwanda and Uganda.
CENTRAL AFRICAN REPUBLIC
SUDAN
0 500km
0 250 miles
N
HAUT-ZAIRE
Zaire (Congo)
EQUATEUR
Kisangani
Bunia
CONGO
GABON
ZAIRE
UGANDA
KIVU
RWANDA
BANDUNDU
EAST KASAI
BURUNDI
Brazzaville
Kinshasa
BAS-ZAIRE
WEST KASAI
Kibombo
Ilebo
Kananga
Mbuji-Mayi
Kabalo
TANZANIA
Lake Tanganyika
Luanda
SHABA
ZAMBIA
Benguela
Lubumbashi
ANGOLA
Kisangani
Zaire
HAUT-ZAIRE
Ubundu
Tingi-Tingi
Bunia
Beni
Kasindi
Ruwenzori
Lake Edward
UGANDA
Kampala
Lake Victoria
Mugunga
Kibumba
Goma
Gisenye
RWANDA
Kigali
Lake Kivu
KIVU
Kalima
Shabunda
Kindu
Bukavu
Cyangugu
Butare
Kamanyola
Mitumba Mts
BURUNDI
Uvira
Bujumbura
TANZANIA
Lake Tanganyika
Kibombo
Kasongo
EAST KASAI
Kalemié was captured by the rebels in early March. It has a hydro-electric power supply and controls a key railway line.
Mbuji-Mayi
Kabalo
Kalemié
SHABA
Moba
0 100km
0 50 miles

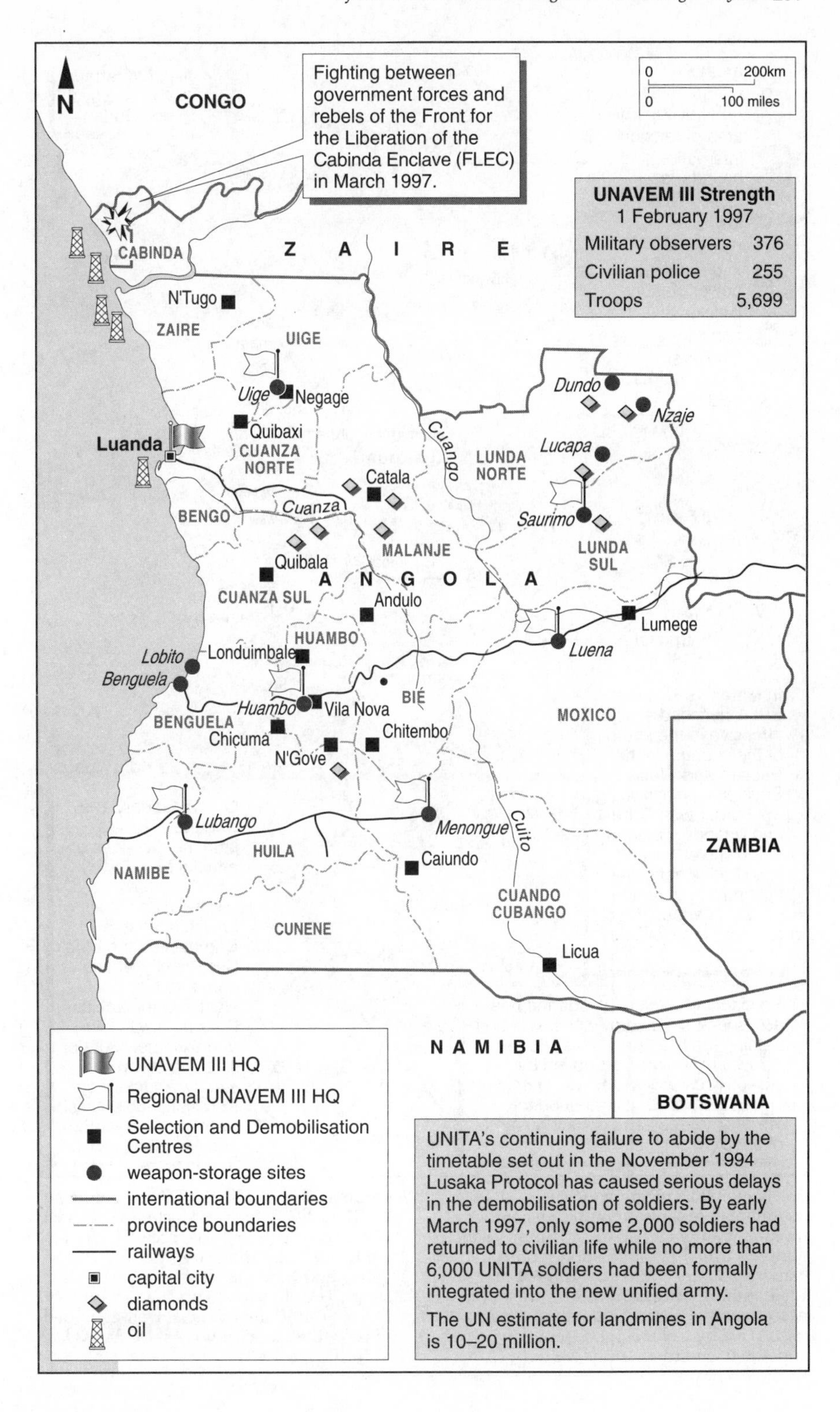
N
CONGO
Fighting between government forces and rebels of the Front for the Liberation of the Cabinda Enclave (FLEC) in March 1997.
0 200km
0 100 miles
UNAVEM III Strength
1 February 1997
Military observers 376
Civilian police 255
Troops 5,699
CABINDA
Z A I R E
N'Tugo
ZAIRE
UIGE
Uige
Negage
Quibaxi
Luanda
CUANZA NORTE
Catala
Cuango
LUNDA NORTE
Dundo
Nzaje
Lucapa
Saurimo
LUNDA SUL
Cuanza
BENGO
MALANJE
Quibala
A N G O L A
CUANZA SUL
Andulo
Lumege
Luena
HUAMBO
Lobito
Londuimbale
Benguela
Huambo
Vila Nova
BIÉ
MOXICO
BENGUELA
Chicuma
Chitembo
N'Gove
Lubango
Menongue
Cuito
HUILA
NAMIBE
Caiundo
ZAMBIA
CUANDO CUBANGO
CUNENE
Licua
NAMIBIA
BOTSWANA
UNAVEM III HQ
Regional UNAVEM III HQ
Selection and Demobilisation Centres
weapon-storage sites
international boundaries
province boundaries
railways
capital city
diamonds
oil
UNITA's continuing failure to abide by the timetable set out in the November 1994 Lusaka Protocol has caused serious delays in the demobilisation of soldiers. By early March 1997, only some 2,000 soldiers had returned to civilian life while no more than 6,000 UNITA soldiers had been formally integrated into the new unified army.
The UN estimate for landmines in Angola is 10–20 million.

The National Democratic Alliance (NDA) is an umbrella organisation uniting southern (predominantly Christian and animist) and northern (Muslim) opposition groups to the National Islamic Front government in Khartoum. The largest of these is the southern-based Sudan People's Liberation Army (SPLA) led by Colonel John Garang who was appointed Chairman of the Military Coordination Council of NDA in September 1996. The NDA, which also includes the Sudan Alliance Forces (SAF), has received support from the neighbouring countries of Eritrea, Uganda and Ethiopia. Its headquarters is in Asmara.The offensives launched in January and March 1997 are the most well-equipped and coordinated operations since the NDA was formed in 1991.

IRAN
Black Sea
RUSSIAN FEDERATION
KAZAKSTAN
Ankara
TURKEY
Aral Sea
N
Caspian Sea
UZBEKISTAN
KYRGYZSTAN
TURKMENISTAN
SYRIA
TAJIKISTAN
CHINA
ISRAEL
Ashkahbad
Baghdad
Tehran
JORDAN
Kabul
IRAQ
IRAN
AFGHANISTAN
Islamabad
Persian Gulf
PAKISTAN
SAUDI ARABIA
INDIA
Riyadh
Red Sea
Arabian Sea
OMAN
YEMEN
Actual range
CSS-8 150km
Scud-B 300km
Scud-C 500km

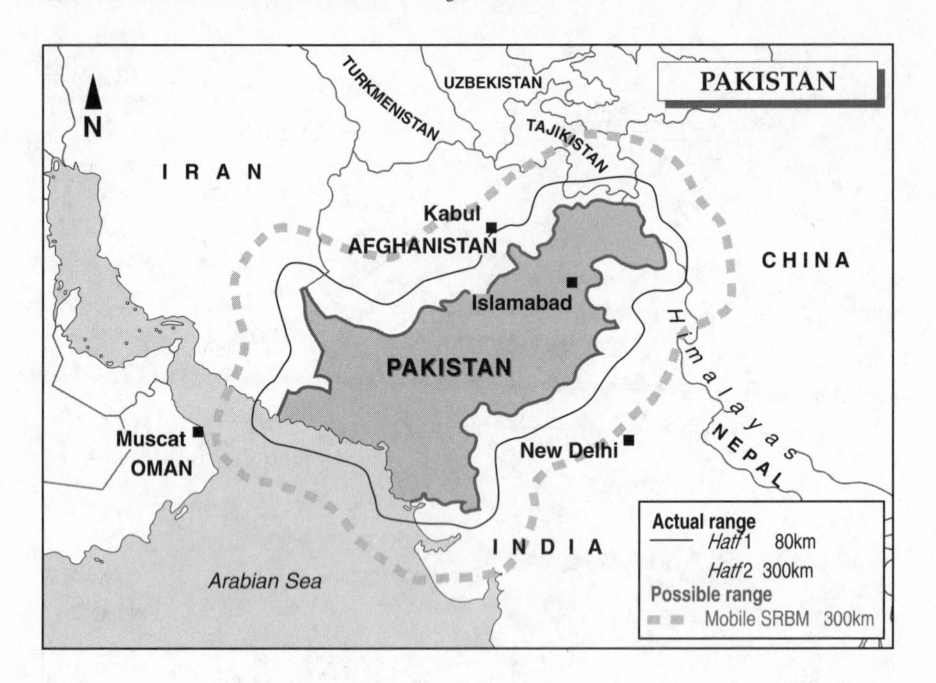

PAKISTAN
N
TURKMENISTAN
UZBEKISTAN
TAJIKISTAN
IRAN
Kabul
AFGHANISTAN
CHINA
Islamabad
Himalayas
PAKISTAN
Muscat
OMAN
New Delhi
NEPAL
INDIA
Arabian Sea
Actual range
Hatf 1 80km
Hatf 2 300km
Possible range
Mobile SRBM 300km

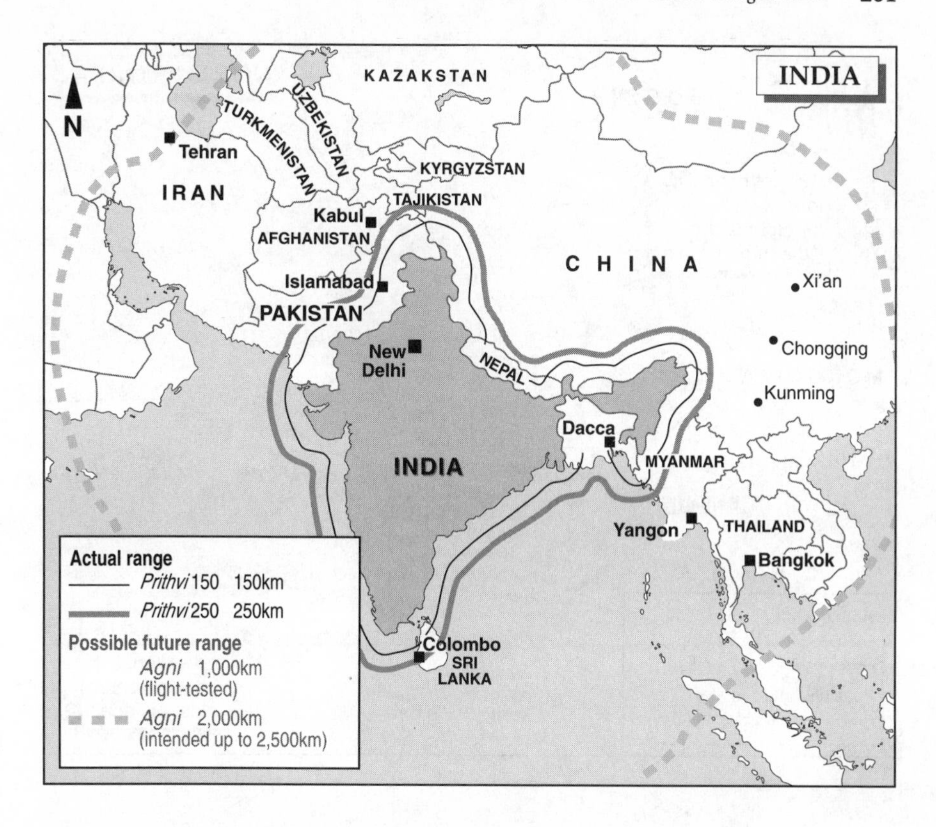
INDIA
N
KAZAKSTAN
TURKMENISTAN
UZBEKISTAN
Tehran
IRAN
KYRGYZSTAN
TAJIKISTAN
Kabul
AFGHANISTAN
CHINA
Xi'an
Islamabad
PAKISTAN
Chongqing
New Delhi
NEPAL
Kunming
Dacca
MYANMAR
INDIA
Yangon
THAILAND
Bangkok
Colombo
SRI LANKA
Actual range
Prithvi 150 150km
Prithvi 250 250km
Possible future range
Agni 1,000km
(flight-tested)
Agni 2,000km
(intended up to 2,500km)

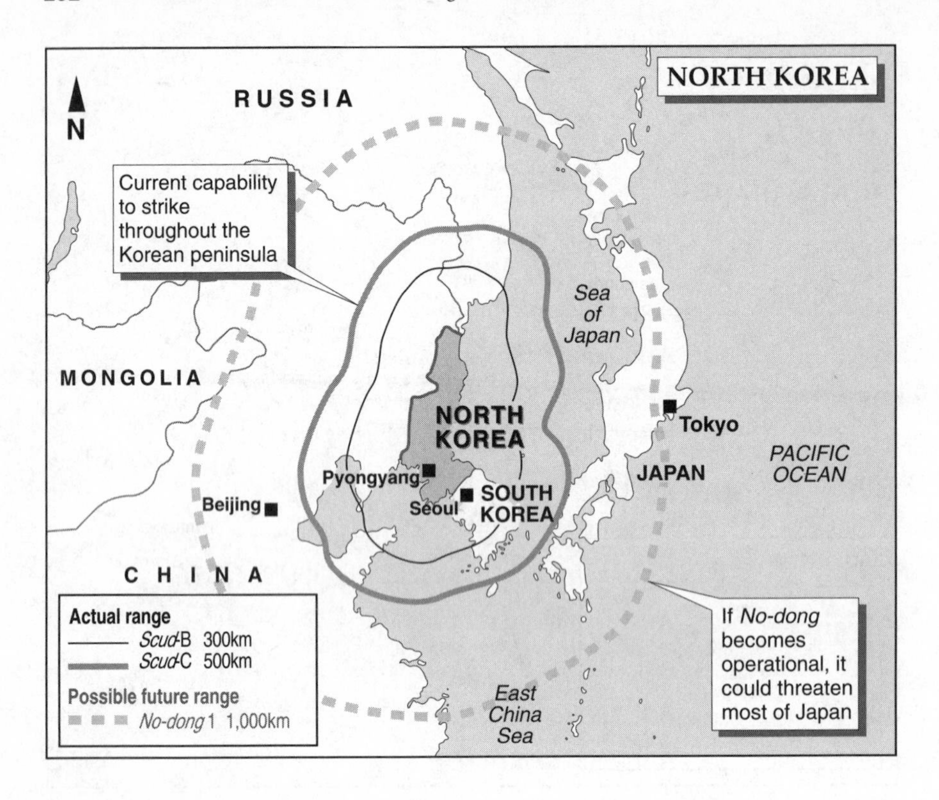
NORTH KOREA
N
RUSSIA
Current capability to strike throughout the Korean peninsula
MONGOLIA
NORTH KOREA
Pyongyang
Seoul
SOUTH KOREA
Beijing
CHINA
Sea of Japan
Tokyo
JAPAN
PACIFIC OCEAN
East China Sea
If No-dong becomes operational, it could threaten most of Japan
Actual range
Scud-B 300km
Scud-C 500km
Possible future range
No-dong 1 1,000km

Glossary

AAM .... Air-to-Air Missile
ABM .... Anti-Ballistic Missile
ACV .... Armoured Combat Vehicle
ADFL .... Alliance of Democratic Forces for the Liberation of Congo-Zaire
ADSAM .... Air-Directed Surface-to-Air Missile
AFV .... Armoured Fighting Vehicle
AID .... Agency for International Development (US)
ANC .... African National Congress
APEC .... Asia-Pacific Economic Cooperation
ARF .... ASEAN Regional Forum
ASCM .... Anti-Ship Cruise Missile
ASEAN .... Association of South-East Asian Nations
ASEM .... Asia–Europe Meeting
Aum Shinrikyo .... Supreme Truth (Japan)
AWACS .... Airborne Warning and Control System
BJP .... *Bharatiya Janata* Party (India)
BMD .... Ballistic Missile Defence
bn .... billion
BTWC .... Biological and Toxin Weapons Convention
BUR .... 'Bottom-Up Review' (US)
BW .... Biological Weapons
BWC .... Biological Weapons Convention
CAR .... Central African Republic
CASOM .... Conventionally Armed Stand-Off Missile
CBO .... Congressional Budget Office (US)
CCB .... Civil Cooperation Bureau (South Africa)

CD .... Conference on Disarmament

CDNS .... Council for Defence and National Security (Pakistan)

CFE .... Conventional Armed Forces in Europe

CFSP .... Common Foreign and Security Policy (EU)

CIA .... Central Intelligence Agency (US)

CIS .... Commonwealth of Independent States

CJTF .... Combined Joint Task Force (NATO)

CNDD .... *Conseil National pour la Défense de la Démocratie* (Burundi)

CPI(M) .... Communist Party of India (Marxist)

CPP .... Cambodian People's Party

CSBM .... Confidence- and Security-Building Measure

CSCAP .... Council for Security and Cooperation in Asia-Pacific

CTBT .... Comprehensive Test Ban Treaty

CWC .... Chemical Weapons Convention

DLP .... Democratic Liberal Party (Republic of Korea)

DM .... Deutschmark

DNA .... deoxyribonucleic acid

DoD .... Department of Defense (US)

DP .... Democratic Party (Japan)

DSMAC .... Digital Scene-Matching Area Correlation

EEZ .... Exclusive Economic Zone

EMU .... Economic and Monetary Union (EU)

EO .... Executive Outcomes

EPR .... *Ejército Popular Revolucionario* (Mexico)

ESDI .... European Security and Defence Identity

EU .... European Union

FBI .... Federal Bureau of Investigation (US)

FELCN .... Special Force for the Fight Against Drugs Trafficking (Bolivia)

FLA .... Future Large Aircraft

FLEC .... Front for the Liberation of the Cabinda Enclave (Zaire)

FY .... Fiscal Year

FYDP .... Future Years' Defense Plan (US)

GAO .... General Accounting Office (US)

GATT .... General Agreement on Tariffs and Trade

GDP .... gross domestic product

GLONASS .... Global Navigation Satellite System

GNP .... gross national product

GNU .... Government of National Unity (South Africa)

GPS .... Global Positioning System

HIV .... human immuno-deficiency virus

IAEA .... International Atomic Energy Agency

ICBM .... Inter-Continental Ballistic Missile

ICJ .... International Court of Justice

ICTY .... International Tribunal for the former Yugoslavia

IEBL .... Inter-Entity Boundary Line (former Yugoslavia)

IFOR .... Implementation Force (Bosnia)

IFP .... *Inkatha* Freedom Party (South Africa)

IGC .... Inter-governmental Conference (EU)

IMF .... International Monetary Fund

JNA .... Yugoslav National Army

juche .... self-reliance (North Korea)

jusen .... housing-loan corporations (Japan)

KDP .... Kurdistan Democratic Party (Iraq)

KEDO .... Korean Peninsula Energy Development Organisation

KT .... kiloton

LACM .... Land-Attack Cruise Missile

LDP .... Liberal Democratic Party (Japan)

m .... million

Mercosur .... Southern Cone Common Market (Latin America)

MFN .... Most Favoured Nation

MITI .... Ministry of International Trade and Industry (Japan)

MOMEP .... Military Observer Mission Ecuador/Peru

MQM .... *Mohajir Quami* Movement (Pakistan)

MRTA .... *Movimiento Revolucionario Túpac Amaru* (Peru)

MTCR .... Missile Technology Control Regime

NAFTA .... North American Free Trade Agreement

NATO .... North Atlantic Treaty Organisation

NDA .... National Democratic Alliance (Sudan)

NFP .... New Frontier Party (*Shinshinto* – Japan)

NGO .... Non-governmental Organisation

NIE .... National Intelligence Estimate (US)

NIF .... National Islamic Front (Sudan)

NP .... National Party (South Africa)

NPT .... Nuclear Non-Proliferation Treaty

NWS .... Nuclear-Weapon States

OAS .... Organisation of American States

OAU .... Organisation of African Unity

OECD .... Organisation for Economic Cooperation and Development

OPCW .... Organisation for the Prohibition of Chemical Weapons

OSCE .... Organisation for Security and Cooperation in Europe

PA .... Palestinian Authority

PAC .... Pan Africanist Congress (South Africa)

PAN .... *Partido Acción Nacional* (Mexico)

PFP .... Partnership for Peace (NATO)

PKK .... Kurdistan Workers' Party (Turkey)

PLAAF .... Peoples' Liberation Army Air Force (China)

PRD .... *Partido de la Revolución Democrática* (Mexico)

PRI .... *Partido Revolucionario Institucional* (Mexico)